Voices in Flight:
RAF Night Operations

Martin W. Bowman

Voices in Flight:
RAF Night Operations

Martin W. Bowman

Pen & Sword
AVIATION

First Published in Great Britain in 2015 by
Pen & Sword Aviation
an imprint of
Pen & Sword Books Ltd
47 Church Street, Barnsley, South Yorkshire S70 2AS

Copyright © Martin W Bowman, 2015
ISBN 9781783831944

A CIP catalogue record for this book is
available from the British Library.

Typeset in 10/12pt Palatino
by GMS Enterprises PE3 8QQ

Printed and bound in England by
CPI Group (UK) Ltd, Croydon, CR0 4YY

Pen & Sword Books Ltd incorporates the Imprints of Pen & Sword
Aviation, Pen & Sword Family History, Pen & Sword Maritime, Pen & Sword
Military, Pen & Sword Discovery, Wharncliffe Local History, Wharncliffe
True Crime, Wharncliffe Transport, Pen & Sword Select, Pen & Sword
Military Classics, Leo Cooper, The Praetorian Press, Remember When,
Seaforth Publishing and Frontline Publishing.

For a complete list of Pen & Sword titles please contact
PEN & SWORD BOOKS LIMITED

47 Church Street, Barnsley, South Yorkshire, S70 2AS, England
E-mail: enquiries@pen-and-sword.co.uk
Website: www.pen-and-sword.co.uk

Contents

Acknowledgements

I am indebted to all the contributors for their words and photographs, in particular Nora Norgate; Jim Moore DFC and The 49 Squadron Association.

Thanks also go to my fellow author, friend and colleague, Graham Simons, for getting the book to press ready standard and for his detailed work on the photographs; to Pen & Sword and in particular, Laura Hirst; and Jon Wilkinson, for his unique jacket design once again.

Prologue

Ghost Fighters Over Dunkirk

D. H. Clarke DFC AFC

In those fateful few days at the end of May and beginning of June 1940 when the British Expeditionary Force lay crammed on the Dunkirk beaches entirely at the mercy of the German Army, Britain could have lost the war. That she did not was due to German lack of foresight and the tremendous efforts made to save the men on the beaches. The soldiers, huddled together on the shell-strafed beaches, were well aware of the efforts being made when they saw the shoals of boats, naval and civilian, queuing to take them off. But, as they were strafed time and time again by enemy aircraft, the cry went up 'Where's the RAF?'

The fact that the RAF were at Dunkirk and played a leading role in the success of the operation is now part of history, although, perhaps, their part was not obvious to the men on the beaches. But I was there too, flying through the misty vastness of the night skies. And if my own aircraft could not be seen by the men below, at least I was in a position to know what happened to the pilots who flew in the daytime.

Despite the fact that Dowding was anxious to husband his front line fighters for what was to be called the Battle of Britain, Group alone maintained average of fifteen Hurricane Squadrons over Dunkirk. But there were also other types capable of carrying guns and it makes not a jot of difference if most of them were antiquated, obsolete or totally unsuitable for the task. These aircraft flew over Dunkirk, apparently unrecognised by our troops - so that in retrospect it seems as if I flew with ghosts.

I operated from Detling, a small aerodrome on the escarpment near Maidstone and normally used by one squadron of Ansons only. I flew there in a yellow and black striped target-towing Blackburn Skua from my base at Gosport on the evening of 31 May 1940 and in company with me was a similarly painted Fairey Battle, also equipped with a D-type winch for target towing and piloted by Pilot Officer Cliff Rendle. In the two aircraft we carried our winch operators. LACs Phelan and Verrier, Sergeant Jefferies as general handyman and Flight Lieutenant Digger Aitken, who was in charge of our little unit. Digger was a regular officer, but Cliff, like myself, held a four-year short service commission. Although I was senior to him by a few months, we had roughly the same amount of hours in our log-books - about 400 - but neither of us had more than two hours night-flying experience. Our total operational experience was nil, so although we were rather surprised at the strange assortment of aircraft which were parked on the aerodrome - Ansons, Battles, a Gauntlet, Vildebeeste, Lysanders, an old Harrow, Swordfish and several other oddments - we certainly did not think of them, or of ourselves as front line fighters.

We asked the way to the Navigation Room and there we found a confusion of pilots and aircrew. Struggling through the crowd, we eventually located the Station CO to whom we had been ordered to report. Group Captain Sainsbury was middle-aged, large and florid: his desk represented an oasis of calm in the surrounding tumult. Digger introduced us and we saluted. 'I understand that one of you has been shown the flares this afternoon,' the Group Captain said. I was startled. It was true that I had been told to fly over to Lee-on-Solent just after lunch 'to have a look at a new idea for target towing at night,' but I couldn't understand how he came to know about it.

'Y-yes, sir,' I stuttered. 'It must have been me.'

'Well, don't you know?'

'I'm sorry, sir.' I pulled myself together. 'I was shown how we could tow lighted flares for night target practice, sir — but nothing was said about coming here to do it.'

'Hm! Night target practice - someone has a sense of humour.' Group Captain Sainsbury accepted the joke, if joke it was, without even a flicker of emotion. Then he went on, addressing us all.

'You are no doubt aware that the British Expeditionary Force has retreated to the French port of Dunkirk just across the Channel - here.' He turned in his chair and stabbed a finger at a four-miles-to-the-inch map pinned to the wall behind him. 'You are probably not aware that the Navy are doing their damnedest to rescue as many soldiers as they can from this untenable position. Now, we at Detling are operating a selection of aircraft day and night to give air-cover and your particular job - code-named 'Flash' - is to illuminate the sea north of the ferry lane so that our bombers can spot any enemy shipping that is around which is likely to interfere with the evacuation.'

The sudden impact of the connection between what I had seen demonstrated only a few hours before and what we were expected to do with the flares, made me feel sick - empty sick: a condition which I soon came to recognise as the twitch and something that was inevitable before any operational flight.

The Group Captain continued. 'You will patrol between Dunkirk and the River Scheldt, about ten miles offshore, lighting your flares one at a time until they are all used. I understand that they burn for 3½ minutes, so take as many as you think you can manage in the rear cockpit - about twenty to thirty should do.' He paused and looked at me intently - then at Cliff. 'You realise, of course, that with 20,000 candlepower to light up the sea, you will also light up yourselves. We don't know if the Germans have any night-fighters in the area, but if they have you will be sitting targets. Consequently I cannot allow you to take maps or documents of any sort with you - in fact, before you go I want you to clear out your pockets completely and check that your crew does the same. Your radio sets should have been removed at Gosport (Digger signified that they had), but I would like you to check through your aircraft and make sure that anything which might be of use to the enemy is removed. We can't be too careful.'

'How about armament, sir?' I asked. 'We're not fitted with guns...'

'I'm afraid that you'll have to do without them. We haven't any here and in any case it would take too long to fit them.' And what is more, he seemed to imply, it would be a waste of time and a waste of guns.

Our first briefing (and brief was an operative word) was a waste of time - apart from the fact that it told us what we were there for. I scarcely had time to show Cliff and our crews how the flares worked, before we were hustled into the air to follow six bomb-laden Swordfish apiece.

I found the night take-off a terrifying business - my previous one being at FTS nearly two years before - and I didn't improve matters by doing it in coarse pitch! Unfortunately, nobody had thought of telling each Swordfish leader to keep his navigation lights on until Cliff and I had pulled into position, so neither of us found their faint blue formation lights and the first trip was a complete fiasco! But we were learning...

In company with fifty other aircrew we managed to doze a miserable four hours on the floor of the ante-room in the Officers' Mess for the rest of that night.

Three nights later I flew on my first operation of the war. Three bombed-up Ansons led the way to Dunkirk and I followed the leader's blue formation lights without difficulty. The weather was perfect, with the barest trace of a mist, but there was practically no horizon and my instrument flying ability was as poor as my experience of flying at night. I had a feeling of unreality - the darkness, no experience, no guns and navigating by guesswork - it was just like a dream. The Perseus droned monotonously and the instruments glowed their luminous messages with green confidence.

Dunkirk was an inferno of fires, surrounded by flashing pin-pricks of light which I assumed were guns hammering ceaselessly at our troops. We turned to port before we reached the land and the leading Anson blinked his formation lights.

'Stream all the wire,' I ordered Phelan over the intercom.

When he reported that the 6,000 feet was out, I pulled clear of the Ansons.

'Right! Let go the first flare.'

Half a minute dragged in agony. I pictured the two-foot tube sliding through the darkness down the long wire. When it reached the toggle at the end the jerk would snap the firing pin home and....

Suddenly the night sky vanished; the faint horizon disappeared. A billion misty droplets of water, almost invisible in the darkness, hurled back the glare of 20,000 candlepower so that I could see nothing outside the cockpit. We were locked in a bowl of brilliant whiteness and it was as if we had flown inside an electric light bulb - even the instruments showed their black and white daytime faces.

Somehow I managed to keep straight and level. When the first flare died, the enveloping blackness which smothered my eyes was even worse than the glare.

'Don't wait for orders - keep 'em lit,' I told Phelan, hoping that he wouldn't notice the tremor in my voice.

For three-quarters of an hour I sweated a blind course up the Belgian and Dutch coast. I never saw a thing - neither land, sea nor Ansons. All I had for guidance were my instruments—how I wished that I had spent more hours under a hood - and some rough courses worked out by the Navigation Officer at Detling.

Then I did see something! A vague blur of movement over the silver disc of the spinning airscrew - half seen through concentrated attention on the instruments. What was it: a night fighter?

There was a sudden jolt. For a moment the engine note changed - then it resumed its steady beat.

'Sir! An aircraft's fouled the wire. The flare's gone!'

And then the blackness... once more the green instruments...

I told Phelan to reel in and he reported that only a few hundred feet of wire were left. Whatever it was I had seen must have flown into the wire and snapped it. With the toggle at the end gone we could light no more flares that night!

I turned on to the reciprocal course. As the pupils of my eyes slowly dilated I began to see again: a faint glow in the sky - Dunkirk; a dark mass over my port wing - land. Thankfully I straightened from my crouch over the instrument panel and flew visually - it was a strain even then, because there was no moon, but with the fires of Dunkirk as my guiding star I found my way back to England.

My orders on approaching the English coast were to fly at 4,000 feet, switch the navigation lights on and off three times and then leave them on and to fire a two-star colour-of-the-day cartridge from the Verey pistol, which was fixed alongside my seat and discharged through the bottom of the fuselage. When I saw the dim outline of home looming ahead I did all these things, confident that was behind us and in my mind I was already despatching a beer...

A white slash of incredible brightness battered into my already overstrained eyes. This time I could see nothing - not even the instruments. I was completely blinded - by our own searchlights! I slammed the seat-positioning lever to the bottom notch, but the low sides of the cockpit, designed for maximum visibility when flying from aircraft carriers, gave me no protection. I flicked the strap-release and strained forward over the stick, my face only inches away from the green messengers of flight. But even in that position the glare was dazzling; the instruments were blank-faced!

I had already fitted a red/yellow cartridge into the Very pistol and I fumbled around until I found the trigger. The discharge thumped. I felt for another cartridge in the rack...

Then, quite suddenly, there was an uncanny silence which even overwhelmed the droning Perseus. A stall! Mentally I felt for the position of the stick and rudder bar. Then we were diving. I closed the throttle as the engine over-revved, easing back on the stick - back, back...She flicked into a spin. I corrected. She went the other way. I got her out. She dived, stalled, spun again and all the time the white finger of blindness followed...And then, miraculously, the lights were switched off; in the few seconds that the carbons glowed, I caught the glimmer of water.

God knows how we escaped! Perhaps we were at 100 feet - maybe less. I corrected the spin instinctively, hauled back on the stick and gave her full throttle... There was a shattering bang! Water showered into the cockpit; the engine vibrated furiously. And then we were clear. We were picked up again by searchlights at nil feet. Some shone down on us so that once again, blinded, I flew without knowing what I was doing or what was happening. I only wanted to land - a ploughed field would have done - anything to escape from the unreality which had turned into a nightmare.

The lights shining down on us switched off; but those behind lit another whiteness ahead - whiteness which stretched up, up... I slammed the throttle open again, dragged back on the stick; we climbed steeply and below, stark in the bleak brilliance, I could see crumbling chalk, tussocks of grass, some sand-bagged pits, white staring faces and then nothingness. The beams were slashed by the lip of the cliff and we were heading inland into blackness...

We had no further trouble from searchlights: I switched my navigation lights off! At zero feet, in the blackout, without maps, after following several wrong creeks, roads and railway lines, I found my way home. I was furiously determined to get there so that I could express an opinion about our reception over the coast; perhaps I would never have achieved it if I hadn't lost my temper. But a temper lost in the panic of danger is soon dissipated by the comforts of safety; and the relief which came when the wheels of my Skua rumbled across the grass between the flaring goosenecks, can only be known by those who themselves have returned.

Late the next day, 3 June, we returned to Gosport - Cliff formating on me as we flew westward into the setting sun. His adventures on the previous night had been very similar to mine, but although we must have been quite close together at times, neither of us had seen the other's flares. Our Ansons, though, claimed to have sunk an E-boat. So what did we care if our odd, tiger-striped aircraft looked a trifle peculiar in formation? We didn't give a damn: we were blooded; we were operational![1]

Endnotes Prologue

1   D. H. Clarke DFC AFC writing in RAF Flying Review, March 1959.

Chapter 1

The Unanswerable Double

'500, 000 pamphlets (weighing one ton) may be more effective than an air raid with 100 tons of explosives.'
Generalleutenant von Metzch, German authority on Total War.

First mentioned in a document by Charlemagne in 794, Frankfurt-am-Main - 'the city of Goethe, the Rothschilds and Anne Frank - is arguably an old city. Standing on the bustling streets of Frankfurt-am-Main in September 1939, one would have hardly noticed the outbreak of war. National Socialism had long prepared German society for the occasion, with air raid exercises, the ubiquitous presence of soldiers in uniform and the sight of marching Hitler Youth all occurrences of daily life since at least the mid-1930s. As German armoured units invaded Poland, the people of Frankfurt expected a tightening of the regime's security measures, a stronger regimentation of their daily routines, more propagandistic grandeur, but no real alteration of their everyday lives. Thus, the first visible signs of war became spectacles of an exciting rather than a menacing nature. RAF aircraft dropped leaflets over the city on 11 September 1939 and Hitler Youth eagerly picked them up and turned them in to their superiors. A few nights later, on 15/16 September, flak fire was heard through the city and on the next day people learned of a RAF bomber shot down over the nearby town of Groß-Gerau.[2] Flying Officer Roland Williams and crew on a 77 Squadron Whitley had been on a sortie to Munich. Williams was killed and his crew were taken into captivity.

In the first week of the war there were five leaflet raids; most of the work was done by Whitleys, but on the night of 8/9 September Wellingtons were out as well. On the morning of 1 September nine Whitley III bombers of 58 Squadron stood ready on the runway at Linton-on-Ouse and began loading a strange cargo. Earlier that day German troops had begun their attack on Poland and general mobilisation had been declared in Great Britain. Bomber crews, listening with bated breath to each news bulletin as it came through on the radio, knew that at any moment they might be called into action. But it was not bombs which were being loaded into their aircraft - or others on 51 Squadron at nearby Leconfield. Instead, parcels of leaflets, each 8½ inches x 4½ inches and headed ominously 'ACHTUNG!' were hurriedly being stacked into the Whitleys. There seemed to be thousands of them; in fact, each aircraft was loaded with over half a million, neatly packaged in bundles of 1,500, each bundle encompassed with a rubber band; twelve bundles making up one package tied round with string. Thirty-two packages were loaded into each aircraft, making a cargo of 1,800lb. After the leaflets had been safely stowed away, the bomber crews sat back and waited for the order to go. It was a strange and unnerving period of suspended animation which was to last two days and nights, while on the Continent Nazi tanks and aircraft rumbled and blazed their way into stricken Poland.

At last, at eleven o'clock on the morning of 3 September, the suspense was over. Prime Minister Chamberlain announced that Britain was at war with Germany and Service stations throughout the country swung into action. At Leconfield, the order came through at five o'clock that afternoon. Three Whitleys on 51 Squadron were to drop leaflets on Hamburg that night; seven more from 58 Squadron were ordered to Bremen and the Ruhr. It was Bomber Command's first operation of the war. At this stage no one knew what sort of reception the bombers would get over Germany. In fact, they met no opposition at all, though the crews reported much searchlight activity around Hamburg. The three Whitleys on 51 Squadron were airborne for only an hour and a half, dropped their leaflets according to plan and returned safely to York without incident. Some of the Ruhr contingent on 58 Squadron however had more difficulty, though not due to the enemy. Engine failure was the chief bugbear; Flight Sergeant Ford in K8990 had to make a crash-landing near St. Quentin on the way back and Flying Officer J. A. 'Tony' O'Neill piloting K8969 'G-George' made a wheels-up landing in a cabbage field at Dormans near Epernay in the heart of the Champagne country. There was a loud thump and the tearing of metal as 'George' skated across the field with cabbages bouncing about the cockpit like 'short-pitched cricket balls' as the broken bomb aiming window sheared them off like 'a harvester's knife'. The crew escaped injury. They had been airborne for seven and three quarter hours. On 5 September a DH Rapide flew O'Neill's crew to Harwell and they were soon back at Leconfield. K8973, the first aircraft to take off, ran out of fuel at 5.45 the following morning and the pilot. Squadron Leader J. J. A. Sutton, had to bring her down at Fécamp. Fortunately, there were no casualties in any of these incidents, but the experience suggested that long distance leaflet-raiding was by no means a piece of cake.

The lesson was apparently learned quickly. For on the following day, when seven aircraft on 51 Squadron were detailed to drop further leaflets on the Ruhr, their load was reduced from 1,800lb to 1,200lb each and they went first to Rheims (100 miles inside France) in the afternoon to refuel. But engine trouble again hampered the operation and only four of the seven were actually able to take off for Germany. Similar troubles afflicted 58 Squadron when they sent four aircraft to Rheims on 9 September. First, the weather was so bad that the operation had to be postponed until the following night; then, when the raid did actually take place, the Whitleys' Mk.VIII Tiger engines played up again. Of the four, only the aircraft piloted by Pilot Officer Sydney Bruce Bintley completed the raid as planned.[3] At this point a quickly changed Government policy brought leaflet operations to a temporary halt. There had been criticism of the propaganda raids almost from the first, for many people were of the opinion that only a good dose of bombs would bring the German leaders to their senses. Indeed, there was a widespread feeling that these leaflets (many of them 'patently idiotic and childish' was the post-war view of Sir Arthur Harris) were doing more harm than good.

The Whitleys operated in the leaflet dropping role on 22 nights between 4/5 September and 23/24 December and a total of 113 sorties were flown. From all these all operations, eleven Whitleys failed to return, the first on 8/9 September when the Whitley piloted by Squadron Leader S. S. Murray was shot down near Kassel and all the crew were taken into captivity. On the same night Flying Officer W. C. G. Cogman on 102 Squadron force-landed at Nivelles aerodrome in Belgium. Flying Officer G. L. Raphael returning from Essen landed at Buc aerodrome in France and taxiing in poor visibility he collided with a parked Dewoitine aircraft. The Whitley was later written off. Cogman's crew meanwhile, were interned for a time and when they were eventually released the Whitley was left behind in Belgium.[4] Two nights' later Sergeant A. G. E. Dixon's crew were injured when their Whitley crashed on take-off from Rheims-Champagne and burst into flames.

Whatever the RAF's own opinion might be of the literary merit of the leaflets there were many in the Air Council who believed in the value of the leaflet raids for reconnaissance and training purposes. Even Harris, who, after the war acidly commented that the only practical result of dropping pamphlets was 'to supply the Continent's requirements of toilet paper,' nevertheless admitted that the operations gave useful training to the crews. Accordingly, on 24 September, the War Cabinet reversed its decision of ten days before and allowed the propaganda sorties to continue. But, in order to placate the 'Give 'em hell' brigade, it was made clear that leaflet-dropping was now only a subsidiary purpose of the operations; primarily they were for reconnaissance and crew training. Later, indeed, leaflets were to be dropped along with more destructive 'packages'. It was this compromise between psychological and physical warfare which was to result in the peculiar air of secrecy which surrounded leaflet operations from then on. Sensitive to criticism that propaganda raids were deflecting Bomber Command from its primary task, the Government was anxious to 'play down' the whole business - at least as far as British ears were concerned. 'Leaflet' became almost a forbidden word: the operations were always referred to in public as 'special reconnaissance' and the code word 'Nickel' was used by bomber crews. The leaflets themselves were closely guarded and file copies were boldly marked in red 'NOT TO BE SHOWN TO THE PRESS.' When an American correspondent asked for copies of the texts to cable to his newspaper, he was told that this was impossible as 'they might fall into German hands'!

It was said that while Poland bled and burned we were bombarding the Germans with nothing more lethal than copies of Mr. Chamberlain's broadcast and the War Cabinet suspended Nickeling operations. With the resumption of operations on the night of 24/25 September, their value in revealing the faults of navigational techniques was at once apparent. This time it was 10 Squadron which provided two Whitley IVs for a 'propaganda prang' on Cuxhaven (euphemistically justified in official records as intended 'to create ARP disturbance in Berlin'). Flying Officer Philips piloting K9033 had no difficulty in reaching the target, but was grossly misled by faulty D/F bearings on the homeward trip, which turned out to be at right angles to true. As a result, he spent most of the night cruising up and down the North Sea parallel to the English coastline and at times came perilously near to infringing the neutral airspace of Holland. Sergeant Verstage in the other Whitley also experienced difficulty with D/F bearings and found his fuel running very low. He managed to land at Leconfield with a few drops still in the tank.

On 1/2 October three Whitleys on 10 Squadron at Dishforth were the first British aircraft of the war to fly over Berlin. Australian Flight Lieutenant John William Allsop and crew were lost without trace. Weather conditions that night were particularly severe. One aircraft arrived over the German capital at 22,500 feet. The oxygen supply momentarily failed; two of the crew collapsed and part of the mechanism of the rear turret froze so that the air gunner could not open his door. The pilot carried on and the navigator went back to assist the two unconscious members of the crew. He dragged one 12 feet along the fuselage into the cabin and connected him with the oxygen supply. He then threw overboard two-thirds of the leaflets before collapsing in his turn. The pilot brought the aircraft down to 9,000 feet and at this height it became possible to open the door of the rear turret. The air gunner climbed through to the assistance of the navigator, who, however, had already recovered and returned to duty. Besides this raid on Berlin leaflets were dropped on eight more occasions in the month of October by aircraft operating mostly from an advanced refuelling base at Villeneuve-les-Vertus near Paris. On the night of 24/25 October when Whitley Vs on 77 Squadron took off from Driffield in Yorkshire on 'Nickel' sorties Pilot Officer Philip Edwin William Walker and crew were lost without trace.

The actual procedure of dropping the pamphlets out of the aircraft seems a simple enough operation when described in cold print. 51 Squadron's historian made it all sound very easy: 'The leaflets are released from the aircraft by first undoing the large packages, which are tied up with string and sliding the bundles so obtained down a specially prepared chute attached to the training flare orifice. As the bundles fall into the slipstream, the rubber band is removed and the leaflets drop singly to earth.'

That things did not work out quite so smoothly in practice was evidenced by the gruelling experiences of five Whitley crews from that squadron when they made a leaflet raid on Stuttgart, Munich and Frankfurt on the night of 27 October. Taking off from Villeneuve shortly after six o'clock in the evening, they ran almost at once into showers of hail and sleet and severe icing. The weather was so bad that one of the aircraft gave up after only 75 minutes' flying and returned to base, but the remaining four kept doggedly on for Germany in spite of appalling conditions. Two were to get back. Leading the operation in K8989 was 51 Squadron's CO, Wing Commander J. Silvester, flying as reserve pilot and observer. The captain was Flying Officer Philip Henry James Budden: the target, Munich. The first hint of trouble was when Sergeant Buckwell in the front turret reported that his gun was frozen solid. Then the trimming tabs froze. But, worst of all, when they were lowering the 'dustbin' mid-turret to drop their leaflets over Munich, the wretched thing jammed and the leaflets had to be hauled across to the other side of the aircraft and pushed down the chute there. All this was done at a height where shortage of oxygen was taxing the crew's strength to the utmost and when they sought relief from their oxygen bottles they discovered that only one was charged. It was necessary to go down and quickly. Fighting off bouts of sickness caused as much by lack of food (none of the crews had eaten since midday) as the bucking aircraft, Budden brought the Whitley down to a less rarefied atmosphere - only to run into heavy anti-aircraft fire. At this point Budden was violently sick and had to hand over the controls to the second pilot, Flying Officer Dermot Evelyn Gould. Then the rear gun and the air speed indicator froze. But somehow they got the aircraft back and following D/F bearings all the way home, touched down at Villeneuve at twenty minutes to one.[5]

The journey of the fourth Whitley, K9008, taking off for Frankfurt ten minutes after the CO, seemed blighted from the start. To begin with, the vacuum pump on the port engine went U/S. Then, after dropping their leaflets the 'dustbin' turret stuck firmly in the down position. Sergeant D. C. Hide and Aircraftman Albert James Heller the wireless operator at length managed to pump it up manually but the effort was so great that Hide fainted. He recovered in time to take control of the aircraft just before the skipper, Flight Sergeant J. H. P. Wynton, collapsed with exhaustion. Then the starboard engine caught fire. It was hurriedly switched off, but the weight of ice on the wings was too much for the now blazing aircraft and it swung down in a steep dive which knocked out Sergeant H. Foster and Corporal Ernest Short in the turrets. With both pilots hauling away at the stick, the Whitley at length came out of its steep dive at about 7,000 feet, but the rudder and elevators were unmovable. The aircraft was now on an even keel but losing height at the rate of 2,000 feet a minute. Wynton gave the order to bail out, but received no reply from the still unconscious gunners. Not prepared to leave them to their fate, he rescinded the order and prepared for a crash landing. The Whitley was now at tree-top level over a French forest and it seemed to the numbed crew that disaster was inevitable. But Hide, who had taken over for the final descent, pulled off an incredible feat of airmanship, skimming over the first belt of trees and bringing the plane down on its belly in a clearing. It travelled through a wire fence, skidded broadside and came to rest with the port wing against some trees. Five dazed, shocked, but otherwise uninjured airmen scrambled from the wreckage and immediately tried to put out the fire in the starboard engine, which was now blazing fiercely. Albert Heller gallantly went back into the aircraft to obtain a fire extinguisher and managed to put out the flames. The crew spent the night in their half-wrecked aircraft and the next morning a Frenchman solicitously asked them what time they would be taking off![6]

An even more miraculous escape was in store for the crew of the last Whitley, K8984, piloted by Sergeant Thomas William Bowles. On the way out ice blanketed the windows and there was snow on the floor of the front turret. Bowles and Sergeant Alan Arnold Emery the second pilot and the rest of the crew kept up their spirits with songs like *'Roll out the Barrel'*, *'Hang out the Washing on the Siegfried Line'* and *'East of the Border'* - a slight geographical adaptation for operations over the Franco-German frontier. But when the 'Nickel' dropping was done the 'dustbin' remained frozen in the down position and the effort to move it manually soon reduced the crew to complete exhaustion. Then the starboard engine gave trouble and near the frontier a cylinder head on the starboard engine blew off. As the Whitley lost height, it descended into thicker and thicker snow clouds and the port engine began to fail. Finally, at 2,000 feet and with hills ahead, Bowles ordered the crew to abandon aircraft. One by one, with varying alarms, they did so, Bowles trimming the aircraft to a slight descending angle before he bailed out. When all this was done the Whitley glided down, bumped heavily and burst into flames and from the rear turret stepped Sergeant A. Griffin the tail-gunner. Blissfully ignorant of the parachute descents - his intercom point had failed at the last moment - he dashed to the front of the burning aircraft to save his comrades. The cockpit was empty. Dazed, cut, burned and more than a trifle puzzled, the sergeant limped his way to the nearest village to a nearby village: here he found the rest of the crew safe in a café, where they exchanged experiences. Sergeant Barber, the front gunner found himself hanging from the aircraft by his neck, caught up in the intercom leads. Frantically, Sergeant Alan Arnold Emery the navigator went to his aid and managed to push Barber clear. Then Emery jumped himself; but having the mistaken idea that he was about to land in water, he began to unloosen his boots on the way down. Terra firma came as a hard shock and he sprained an ankle. Barber was knocked unconscious by his parachute which, when opening, hit him on the head. He regained consciousness lying on his back in a field among a herd of curious but friendly cows. Aircraftman R. Jackson, the wireless operator, after sending out a final message and clamping down the key, found he had a problem on his hands - literally. His oxygen bottle was frozen to his fingers so, when his turn came to jump, the oxygen bottle went with him. He landed in a singularly inhospitable field, whose occupant, an over-excited bull, resented the intrusion in no uncertain manner. Fortunately, Jackson was able to sprint to the safety of a hedge. Landing in a field full of curious but hostile bulls, he took successful avoiding action by sprinting for the hedge in full flying kit and cleared a four-foot hurdle. Bowles trimmed the aircraft in a shallow dive and bailed out. He landed softly and was unhurt. The whole crew after their reunion in the café, were taken, 'complete with bouquets,' to a French hospital, whence, after treatment, they returned the same day to their Squadron.[7]

The only flowers for Squadron Leader James Aldo Bartlett Begg's crew on 77 Squadron following the 'Nickeling' sortie from Villeneuve on 10/11 November were at their funeral at Charmes Military Cemetery after their Whitley crashed at Bouxurulles killing everyone on board the aircraft.

The most interesting development in the following months was the publication of a miniature newspaper called *Wolkiger Beobachter* (*Observer from the Clouds*), the first issue of which was dropped on 20/21 November. Only thirteen issues in all were produced between then and the following April, but it provided valuable experience for the later 'Nachrichten', of which a million and a half copies were dropped over Germany every day between April 1944 and May 1945.

Extreme weather conditions over the Continent halted all propaganda raids during late January and early February, but when they were renewed, the aircraft of Bomber Command found themselves going further afield, to Prague, Vienna and even Warsaw. By 6 April 1940 when leaflet operations over Germany - except by the Advanced Air Striking Force - were suspended, Bomber Command had delivered 65,000,000 leaflets, or nearly 9,000 miles of paper. The accuracy and determination displayed during these operations was a significant pointer to the more lethal attacks which were to follow in the years ahead.[8]

On 17/18 May 48 Hampdens went to Hamburg and 24 Whitleys to Bremen, all attacking oil refineries and six Wellingtons visited the railway yards at Cologne. There were no bomber losses but German casualties were beginning to mount. Forty-seven people were killed and 127 injured in Hamburg and Bremen. In Hamburg there were 36 separate fires, the largest causing the complete gutting of the Mercksche fertilizer factory and 160 buildings damaged by high-explosive bombs. The most seriously damaged building after the fertilizer factory was in the Repeerbahn. In all, thirty-four people - (twenty-four men, nine women and one child) - were killed and 72 individuals had suffered injuries of varying degree. Hans Brunswig, a senior officer of the Hamburg Fire Brigade, describes the attack on his city:

'In the meanwhile, all hell is let loose outside: over the roofs of Harburg streak tracer trajectories from the 2 cm flak. Dull detonations can be heard and the 8.8 cm flak battery at the access road to the autobahn at Harburg is firing salvos, other heavy batteries near and far joining in. In the sky there is the cracking and the flashing of detonating flak shells - and on the ground, too, of impacting high-explosive bombs and dazzling-white incendiaries. Numerous parachute flares turn night into day. In rather hazy weather they had dropped about eighty high-explosive bombs and 400 incendiaries on targets lit by parachute flares, mainly in the Harburg industrial area and in doing so had caused, in addition to considerable high-explosive damage, six major fires spreading over a considerable area, one medium-sized fire and twenty-nine minor fires.'[9]

When the first bombs hit Frankfurt in June 1940 killing five in the suburb of Nied and thirteen in Höchst just west of Frankfurt, the event was overshadowed by the euphoria over the Wehrmacht's astonishing victory over France. As elsewhere in Germany, the people of Frankfurt saw the end of the war impending, with Germany victorious and little loss to their own home city. Herbert Heckmann, then ten years old, noticed the signs of optimism. 'At the beginning of the war there was much chatting in our street. Letters from the front were discussed and the news of the victories, which came barking out of our small, black radios, enticed some to make daredevil speeches, so that war veterans who had experienced defeat in World War One now could hardly be restrained heading to the front in order to be on the side of the victors this time.' Frankfurt's main threat, of course, came from aerial attack. Air raid preparation, which had begun as early as 1934, had led to the construction of several hundred shelters across the city. Throughout 1940 almost fifty larger bunkers and several make-shift underground hospitals were added, turning the city into one of the best prepared in Germany. Nightly air raid alarms became a regular occurrence. Even though RAF bombers appeared often, the number of air raid casualties in Frankfurt remained low. Bombs killed 31 people in 1940 and thirty in 1941. While the numbers 1940 and 1041 were hardly surprising as the RAF struggled to bring the air war to Germany's interior, the number for 1942 would be less so. [10]

In Britain the attitude was 'keep smiling through', 'keep calm and carry on'. Tom Imrie, a RAF rear-gunner, recalled: 'After completing my training on Whitley Vs at 10 OTU, Abingdon, as a Wireless Operator/Air Gunner, I reported to 51 Squadron, RAF Dishforth, on 1 September 1940. I was assigned to rear turret duties for my first night operation on 9 September to bomb the naval base at Bremen. Those who knew the rear turret on the Whitley with its four Browning .303s had been suitably impressed that it was essential to make sure the double doors were securely locked before the turret was operated. This required the pulling of one door towards your back by a strap and, having got yourself into position, closing and locking the other door with your spare hand. On this first trip all seemed well until the 'skipper' ordered gun testing and turret movement. I dutifully swung my turret through 45 degrees to be greeted by a loud bang - a door had opened and I was jammed with my backside out over the North Sea and my parachute out of reach in the fuselage. An unpleasant couple of hours followed while we bombed secondary targets on the Dutch coast and returned to base where I was suitably chastised. Not a good way to start one's squadron service, but perhaps a blessing in disguise. I was, from that night on, confined to the comparative comfort of the wireless operator's position adjacent to the navigator.' [11]

Australian pilot David Shannon flew Whitleys at OTU Kinloss. Some short while after his 18th birthday in 1940 he and his 'great chum' Batty Marks, presented themselves at the Royal Australian Navy's recruitment office for volunteers in Adelaide. When they were told that they could expect a long wait before being called up for training, Shannon walked around the corner and signed on with the RAAF to train as aircrew. Batty awaited his call from the Navy. He was to spend a great deal of his war service in Arctic waters protecting the Russian convoys. Shannon graduated as a pilot in single- and twin-engined aircraft and was shipped off to England. It was his destiny to be seconded to RAF Bomber Command. On his first night solo in a Whitley doing circuits and bumps he had an engine cut at about 200 feet, just after take-off. 'I did a great wide right-hand circuit and got back on the runway' he recalled. 'The comment of my flying instructor, back in the crew room, was forthright: 'We never expected to see you back in here; coffins don't fly so well on one engine!' [12]

Even as early as the night of 1/2 October in freezing weather, Whitleys went from their base in England to Berlin. Flying at a great height, the crews depended entirely on oxygen to keep them conscious; when, at over 20,000 feet above Berlin, the oxygen supply in one Whitley failed for a moment two of the crew at once collapsed; they revived only when the pilot had dived down to 9,000 feet. This roused the anti-aircraft guns but the guns missed and the Whitley was soon in the upper air again. There was every sign that the inoffensive leaflet raids were doing nothing to check the offensive spirit of the crews. No doubt there was some feeling among the crews, as there was among the general public, that their own extraordinary exertions were being wasted in a foolish paper war. But they never suffered from that deadly feeling that the whole war was unreal.

On a single flight one bomber came first into an electric storm and then into a snowstorm and then at 20,000 feet, ice formed on the airscrew. The pilot took off his glove for one moment to adjust his controls and in that moment he was frostbitten. As the winter went on the crews found that they had yet more to learn about cold. Ice formed actually within the bombers at 8,000 feet and even at 1,000 feet masses of ice sometimes settled on the wings. When one bomber on the way to Munich was at 10,000 feet the temperature was 28°C below zero and ice completely covered the front windows of the cabin. The wings were loaded with ice and the crew could hear chunks of ice coming off the airscrew and hitting the sides of the bomber's nose. The controls had to be kept in movement all the time to prevent them from freezing stiff. The pain from cold and frostbite was so acute that some of the crew, longing for anything else but that, even any pain of another kind, butted their heads on the floor and on the navigation table. The log of a wireless operator on a Wellington reconnoitering north-west Germany in February 1940 recorded that for the first two hours of the flight he was using oxygen at 17,000 feet and once or twice wrote: 'All quiet.' Then he wrote: 'Electrical storm. Both aerials earthed. Aircraft iced up. Going down lower.' Next: 'Heater U/S' (unserviceable). On the way back he wrote: 'Still over enemy territory'; and two minutes later: 'Crossing enemy coast.' Over the sea the log recorded: 'Feet freezing. Can't feel sense in feet. Breathing not too bad. Induces sleep. Eyelids covered in frost.' Then he referred to his wireless apparatus: 'Volume control and reaction controls frozen.'

Meanwhile the defences of Germany were getting more efficient. Night fighters were not yet a serious menace, though some were seen. On 22/23 March 1940 a night-flying Messerschmitt fired for the first time at a Hampden engaged on reconnaissance of north-west Germany. The Messerschmitt 110 came out of cloud and approached the Hampden from behind; it made its attack from the starboard quarter. The Hampden's rear gunner had time for one burst and then the pilot made a spiral dive towards a cloud-bank and so got away. After the Hampden had landed at base a few bullet-holes were found in it; they were near the first-aid pack. On the same night bombers over the Ruhr reported night fighters, but no attack. On the same night there was reported an increase of anti-aircraft fire and of searchlights. At intervals during the winter the guns were getting thicker on the ground and working in closer co-operation with searchlights. The Germans were using projectiles of all kinds: shrapnel, 'flaming onions' - a string of balls of fire shot into the air in the presumed course of the aircraft - and incendiary shells, 'which on exploding' - to quote a report made at the time - 'discharged large balls of fire that could be seen the whole way up. They eventually disappeared without bursting.'

On the night of 31 July/1 August 1940 42 Battles, Blenheims and Hampdens set out to bomb a variety of targets and sow mines in enemy waters but only thirteen bombed and nine laid mines. On its outward flight, one Battle was shot down into the sea off Skegness by RAF fighters and three Hampdens ditched in the sea on the return. Squadron Leader R. L. Oxley DFC, who was involved in the 'Gardening' operation, got his Hampden safely back to Lindholme but overstressed the aircraft. 'Oxley, who was known among his service friends as 'Beetle' showed as much effrontery as his namesake in Kipling's Stalky and Co., recalled David Masters in *So Few: The Immortal Record of the RAF'* published in 1941 'with results that surprised him - as well as the enemy. It started with his bomber acting as a decoy over Magdeburg [probably on the night of 3/4 September], making as much noise as possible in order to draw the fire and let his companions go in and do their work before he went off to do his. Low cloud baffled the searchlights and Squadron Leader Oxley was not worried by the guns, which were firing blindly. Having roared about over Magdeburg to scare the Germans, he climbed to 6,000 feet to bomb his target which had shown up for a moment through a gap in the clouds. Swiftly, unexpectedly, the sinister shape of a German balloon threatened death only fifty yards away. It gave him a shock, 'I thought of the chances I'd been running, flying in and out of those cables for half an hour,' he said later. 'As I turned and climbed I saw five or six dirty, grey shapes, just like the British balloons.' Incidentally he once saw a German balloon as high as 13,000 feet. Dodging the balloons, he dropped a bomb or two on his target and turned to find another target where he could drop the rest. Flying south-west of Bremen at 10,000 feet, he saw an enemy aircraft landing on the illuminated flare path of an aerodrome. The red lamp was left on and the captain of the bomber was not slow to seize his opportunity. Gliding down to 4,000 feet, he fired off a signal cartridge to see whether the enemy would put the lights on. The Germans fired up a signal consisting of three yellow and three white lights in reply.

'The nearest thing I had that night was a red-yellow, so I fired that,' recorded Squadron Leader Oxley. 'I was quite enjoying myself to see if we could fix them. They fired another signal, so I switched the navigation lights on, but these had no effect. Then I started Morsing on the recognition lights and for want of anything better to send I sent 'Heil Hitler.' That had the pleasing and astonishing effect of getting them to put all the lights on. They gave me a green to come in and land, which was just what I wanted, I went round the circuit as though to land-and opened up and let our bombs go into the hangars. Then I went off and returned about twenty minutes later. They must have been convinced that no enemy would hang about so long, for when I again signalled with my recognition lights, they gave us another green to land. But we had nothing more to drop. I wonder what they told the CO next day!'

'Which incident goes far toward explaining his nickname. Yet there is nothing of the dare-devil about his appearance. He looks far too good-natured to go bull-baiting Germans in their own arenas. He speaks quietly and moves with an easy step. About five feet five inches in height, he is inclined to be tubby, with a clean-shaven face and fair hair that is going thin on the top, though he is still in his twenties. It is when his vivid blue eyes light up with a mischievous twinkle that he betrays a hint of that sense of humour which has found such free expression over Germany, although the Germans will never see his jokes.

'Like other pilots, he struck a good blow for England when Hitler was preparing to invade Great Britain during the, crisis of 1940 and Squadron Leader Oxley twice ran the gauntlet of the heavy enemy defences in order to bomb the barges being collected for the transport of German troops. The Germans had mounted large numbers of machine-guns controlled electrically from a distance and during his first attack Squadron Leader Oxley was so blinded by the searchlights that he had to keep his eyes on the instrument panel and steer as instructed by the navigator.

'A few nights later he arrived at the same spot at 2202 and dropped a flare. 'Oh my God sir, you can't miss. This is wonderful!' exclaimed Sergeant Horner, the navigator. There were about a hundred barges arranged in arrowhead formation along the shore. The flare set all the guns firing in the neighbourhood as Squadron Leader Oxley circled to make a dive-bombing attack. At 5,000 feet the searchlights picked up the bomber and made it impossible for the pilot to see, so he flew once more under the instructions of the navigator. 'Push her down a bit, sir!' the latter exclaimed when they were down to about 3,000 feet. A moment later he added: 'Bombs gone!'

'What do we do now?' asked the captain.

'Keep on going down' came the reply, as the captain closed the bomb doors. 'Now turn left,' said the navigator when they were down to 300 feet.

Escaping the dazzling lights, the captain dropped still lower, leaping a sand dune like a horse going over a hurdle at Aintree and climbing out to sea for home. His bombs shattered the barges, covered with black tarpaulins, but none of the bomber crew saw any German troops on board, which was not surprising considering that it was at night. The bomber's attack lasted exactly four minutes.' [13]

Throughout August, just when one would expect the story to come to a climax and the threat of the flat-bottomed boats to declare itself, there was a long pause. In daylight Blenheims were almost wholly engaged in attacks on aerodromes and the long-range guns at Cape Gris Nez. These were bombed on many nights from the middle of August onwards and well on into September. As to the barges, they were still about on inland waterways, but they had not, as yet, collected in the Channel ports in any great quantity. In August 1940 traffic on the Dortmund-Ems Canal, an essential link between central and western Germany and a bottleneck of the canal system, was about three times what it had been before. As a result, bombing attacks were made on this canal, highly vulnerable where it is carried across aqueducts, at frequent intervals. German transport had to be attacked whatever might be under the covers of those mysterious barges; this was the period when Bomber Command made Hamm a household word and traffic diverted from much-bombed railways would naturally go by canal. For the time being, with invasion apparently not absolutely imminent, it probably seemed better to bomb where an attack would be a threefold gain, helping to interfere with Germany's economic organization, hampering the supply of armies in the West and blocking the route for invasion barges coming from the East.

Hampden pilot, Wilf Burnett, recalled: 'The Station Commander gathered all officers together one morning in August or September 1940 and told us that it appeared invasion was imminent and that we should be prepared for it. I remember the silence that followed. We left the room and I don't think anyone spoke, but we were all the more determined to make certain that we did everything possible to deter the Germans from launching their invasion. At the time we were bombing the invasion barges in the Channel ports, undertaking operations almost every other night. I remember one operation in particular against the invasion barges. We had part moonlight, which was very helpful because navigation in those days depended entirely on visual identification. We flew to the north of our target so that we could get a better outline of the coast. We followed the coast down towards our target, letting down to about 4,000 feet so that we could get a better view of what was below and to increase the accuracy of the bombing. At that height light anti-aircraft fire was pretty heavy and fairly accurate so we didn't hang around after dropping our bombs. This was done repeatedly over a period of time until the invasion was called off.'

Twenty-seven year old Flight Lieutenant Roderick Alastair Brook 'Babe' Learoyd, a pilot on 49 Squadron at Scampton, was awarded the Victoria Cross for a most daring low-level swoop on one of the Dortmund-Ems Canal's aqueducts on the night of 12/13 August. Learoyd recalled later in a BBC broadcast: 'Our target on this raid was the old aqueduct carrying the Dortmund-Ems Canal over the River Ems north of Münster. This canal is of great importance to the industrial area of the Ruhr. There is also at this point a new aqueduct, but when that was blown up as a result of previous raids the Germans had diverted all traffic to the old one. The operation had been most carefully planned. Five aircraft detailed for bombing, were to slip in and carry out their work. Two of the five, I am sorry to say, never got back.

'Timing was an all-important factor. For a reason I cannot mention it was imperative that the five of us should all attack within a very short period.

'At three o'clock in the afternoon we were told that we were going and at six o'clock that evening we were given the details of the operation. Aircraft from two squadrons were taking part. Having been there before, most of us knew the place pretty well. The actual briefing of the crews took about three-quarters of an hour. The whole place was carefully gone through with special maps and plans.

'We synchronised our watches and the clocks in the aircraft before starting. Everybody got away right on time. Just after we took off, I saw one of the others in the air, but we soon lost sight of him. The timing had been worked out so as to allow us a ten-minute margin in case we got slightly off our course or had any trouble in getting into the target area. My navigator did a very fine job of work and we arrived at a point north of the target with our ten minutes in hand, so we circled round there for a bit.

'Going out, there hadn't been any excitement, but we were not looking for trouble anyway. There were clouds on the way over but they cleared beautifully just on the edge of the target. The moon was about half full. We were relying on the moonlight reflecting on the water to give us our direction for the run up. We being the last of the five were due to go in at 23.23. Two minutes before that time we came down to about 300 feet. We were then still several miles north of the target. Gradually we lost height as we came along the Canal, following its course all the time.

The navigator was in the nose of the aircraft doing the bomb-aiming. Everything was quiet until we got to the point where the Canal forked just before the two aqueducts. I was doing the run up to this point when the navigator was taking over the directing. We must have gone off a bit to the left because he called out 'Right', then immediately after, when we had turned a bit to make the correction, he called out 'Steady'.

'Then, suddenly, everything started at once-searchlights and all the anti-aircraft fire. It was unfortunate from our point of view of course, that the enemy knew pretty well the direction from which we must attack. They had disposed their defences so that they formed a sort of lane through which we had to pass. It seemed to me that they had strengthened these defences a great deal since the first raids.

'The searchlights were blinding and we were flying entirely on the bomb aimer's instructions. I had my head down inside the cockpit trying to see the instruments, but the glare made even that difficult. Our instructions were not to rush it too much because of the need for extreme accuracy. Before we started, the rear gunner had asked if he could fire at something or somebody and he was shooting at the searchlights as we went past.

'Almost at the same moment as we bombed I felt a thump and the aircraft lurched to the right. A pom-pom shell had gone through the starboard wing. Then another shell hit the same wing between the fuselage and the engine. They were firing pretty well at point-blank range. It was all over in a few seconds. The navigator called out 'OK finish'. Then we turned away again. The ground defences were still after us but the tracer was dying out a bit by this time.

'When we had got away and set course for the base the rear- gunner reported that oil was coming into his cockpit. Then the wireless operator reported that the flaps were drooping. I tried to raise them but found that they wouldn't come up. What had happened was that the hydraulic system had been damaged. We discovered too that the undercarriage indicators were out of action.

'Not having landed without flaps before I didn't like to try it that night with a crew aboard, so we cruised around a bit doing a few local 'cross countries' for about two and a half hours. We waited till dawn and then we came in all right.'

It was in an attack on Antwerp on Saturday, 15 September 1940 that Sergeant John Hannah's Hampden bomber on 83 Squadron at Scampton caught fire and that he was awarded the VC for extinguishing the fire. The blue-eyed, fair haired 18-year old Scottish wireless operator/air gunner was on a makeshift crew of Hampden P1355 'Betty' flown by Canadian Pilot Officer Clare Arthur Connor. Hannah, born at Paisley in November 1921 had served as a salesman in a Glasgow shoe shop before he joined the RAF in 1939. The navigator was Flight Sergeant Douglas Hayhurst recalled from three weeks' leave, his reward for flying 38 ops on many months of night operations against the enemy. Sergeant George 'Buffy' James, the rear gunner, had flown nine ops. Connor, who was awarded the DFC, later told their story in a BBC broadcast:

'In order that you may get a clear picture of what happened that night, when my aircraft caught fire and when Sergeant Hannah performed that very fine act of bravery which earned for him the Victoria Cross, it would, I think, help matters if I described to you the interior of our Hampden bomber. In the nose of the machine sits the navigator. He is the most comfortably placed member of the crew. He can almost stand up and has plenty of leg room. I, as pilot, sit in another cockpit behind and above him and from the moment the aircraft takes off until it returns I do not leave my position. The other two members of the crew, who ate the rear gunner and the wireless operator, occupy the top and bottom gun positions. They can however, if they wish, crawl through the aeroplane from one end to the other.

'I have gone into these details so that you may, the more easily, realise that Sergeant Hannah had not only to contend with the raging flames but was called upon to extinguish them without at any time being able to pull himself up to his full height. It must be remembered too that he was wearing full flying kit which saved him from serious bodily burns but at the same time restricted his movements. Before I say anything more about Sergeant Hannah I want to describe, as best I can, what happened that night. If anybody had told me that only half the crew and three-quarters of the aeroplane would return to England I should have been inclined to laugh at them - but that's what happened.

'We left in fine weather in high clouds and in due course we were over Antwerp. We started to make our bombing run but found that we were not in line to make a good attack, so we turned, circled round and got into better position. As soon as we arrived we noticed that the anti-aircraft gunfire was fairly heavy, but during that first run none of it came very close to us. It wasn't long, however, before they got our range and as we came round for the second attack we met a terrific barrage. We were hit in the wing on the way down several times and the aircraft shook so much that it was not an easy matter to keep control of it. However, we released our bombs and it was then that I saw flames reflected in my perspex windscreen, but I was so busy taking violent evasive action against the anti-aircraft guns that I didn't at first give it any serious thought. Whilst I was avoiding the shells - as best I could - the wireless operator called me on the intercommunication system and announced, very quietly, in his marked Scots accent, 'The aircraft is on fire'. I asked him 'Is it very bad?' He replied, 'Bad, but not too bad'. I gathered from this conversation and from the fact that the reflection of the flames was getting brighter and brighter, that the position was fairly serious. Sergeant Hannah, cool as he was, did not want to alarm me. I immediately warned the crew to prepare to abandon the aircraft, at the same time I was still throwing the machine all over the place in an effort to dodge the shells some of which were ripping right through the fuselage and others seemed to be bouncing off. Besides this heavy stuff there was a lot of tracer shooting all round us and I was not very keen on my crew jumping through that. Their chance of landing unharmed would have been small.

'In the meantime the fire was getting an even firmer hold and I imagine that the blazing aircraft must have presented the enemy gunners with a pretty good target. After three or four minutes of more shells whizzing through us and past us I was relieved to find that we were at last out of range and I think it must have been about this time that my navigator and rear-gunner jumped for it. There is no doubt that the navigator was quite convinced that there was no chance of the aircraft surviving, whilst the rear-gunner apparently had no option. He was literally burned out of his bottom cockpit in circumstances which must have made it impossible for him to stay there.

'The fact that the rear gunner did jump gave Sergeant Hannah more freedom of movement. Whilst he was fighting the flames with his log book and with his hands I could feel the heat getting nearer and nearer to the back of my neck, but at the same time I noticed, when I turned round, that the flames were still some four or five feet away from me. At first Hannah was wearing his oxygen mask, but the fumes were evidently too strong and he found himself beginning to suffocate. So, without any hesitation, he ripped the mask off and dashed through the fire heedless of the burns which he could not possibly avoid. After about ten minutes, which seemed like hours, I noticed the reflection in the windscreen had died down and that in place of the heat at the back of my neck there was a welcome and refreshingly cool breeze. I asked the sergeant on the intercommunication system, which miraculously escaped damage, how things were going. He said, in his cheery manner, 'The fire is out, sir'. I then asked him how the other members of the crew were getting on. He said, 'I'll find out, sir'. He then went into the rear-gunner's cockpit and said, 'Nobody here, sir'. He then climbed forward to the navigator's position and reported 'Navigator not present. We are all alone, sir'. He then scrambled into my cockpit and brought me the navigator's maps so that I could steer a course for home. I realised what he had gone through. His face was badly burnt, his flying suit was scorched all over and altogether he looked a sorry sight. Through it all he was grinning and I then knew that although his injuries were severe they were not as bad as they looked. On the way back home Hannah sat in the navigator's position away from the smell of the fire and when we landed he jumped out of the aeroplane as though what he had done had been an everyday occurrence. When I looked at the machine I got some idea of what he had gone through. The rear-gunner's cockpit and half the interior of the fuselage were charred ruins. There was a hole in the fuselage large enough for a man to crawl through. There were holes in the wings, but far more serious were the holes in the petrol tanks and how the petrol didn't catch alight and undo all Sergeant Hannah's good work will remain a mystery. I believed that Hannah was fully conscious of that danger and concentrated on the flames nearest the tanks before he dealt with the other fires which broke out. To make matters even worse, while he was beating out the flames, thousands of rounds of ammunition were going off in all directions and he had to fight his way through this fierce internal barrage to save the aircraft. He didn't give his own safety a thought. He could have jumped, but preferred to stay behind.'

Attacks like these have brought about a hundred adventures; here is one of them. On the night of 11/12 September the pilot of a Wellington broke cloud just south of Flushing. His target was the docks and for twenty minutes he flew round to make sure of his position. As soon as he had pin-pointed his target he turned out to sea and then glided in across the docks. The enemy were ready for him; the searchlights caught him and up came the flak. 'There wasn't,' the pilot said afterwards, 'a single hole in the barrage that we could have got through. It didn't seem worth trying to dodge the shells. I thought, 'Good night, nurse'; I put the nose down and hoped for the best.'

The Wellington went down very fast. Inside there was a reek of cordite. But nobody said anything. It flashed across the pilot's mind that if the Wellington was brought down at least it would take its bombs down to the docks with it. The pilot looked at the docks, looked at his instruments, looked again at the docks. At 150 feet the bombs were dropped. As they burst on dock buildings their concussion shot the Wellington up to 600 feet. The second pilot was standing up; his knees buckled beneath him and he was thrown to the floor. As to the pilot - 'More or less automatically,' he said, 'I pushed the nose down, the throttles forward and the governors up. Shells were bursting on every side. We went through the western side. That was our way home.' Then the flak ships round the harbour joined in. They used heavy quick-firing guns and shot up 'flaming onions' by the dozen. Tracer from machine-guns spiralled in the air. The Wellington tore along the edge of the coast while the crew looked down at the white tops of the breakers. The crew knew that their bomber was hit; the pilot had in fact felt fragments of shell smack into it. The wireless operator signalled to base to say that they would try to make England, but might be forced down on the water. For many minutes they could see, away behind them, the bright bursts of explosions in the Flushing docks. The intercom to the rear turret failed, but the pilot did not realize this at once; he tried to speak to the rear gunner and when there was no answer he thought that the gunner might be wounded or killed. The front gunner came out of his turret to get ready the dinghy gear, a task which properly belongs to the rear gunner, in case it should, be wanted. All the while the Wellington was kicking like a bronco and there was heavy cloud and rain all the way to England. But the pilot at last brought the aircraft home and every one of the crew safe with it.

One evening in August 1940 31-year old Pilot Officer Richard C. Rivaz, a tail gunner who when he had volunteered for the RAF found that he was too old for pilot training, arrived on 102 Squadron flying Whitleys at Driffield 'feeling very new and shy and rather wondering what sort of people he should meet and how they would treat a new boy like himself.'

'The only operational crews I had seen was when an odd crew had landed at my OTU base on their way home after a raid. These people had always been dressed in flying boots and were wearing no collars or ties, but had silk scarves knotted round their necks and they were usually unshaved and with un-brushed hair. I had looked on them as some sort of gods and wondered whether one day I, too, should be privileged to walk about and look as they did. These were the people I should be meeting now and with whom I should have to live. Somehow they did not seem to me to be ordinary normal people, but people either with charmed lives or else lives that would soon not be theirs and I thought this would surely be visible in either their appearance or behaviour.

'I was quite surprised to find that the Officers' Mess was very similar to the one I had just left. I arrived at Driffield after supper and found my way to the ante-room, where the wireless was on, apparently unnoticed by anyone in the room. There were some people lolling in deep black leather arm-chairs, reading and one or two were asleep. There was a group standing round the empty fireplace with pint beer-tankards in their hands. Some were writing letters and four were playing cards at a table in the middle of the room. Everyone there looked perfectly normal; in fact, the whole scene; as I surveyed it was just the same as could be seen in the ante-room of the Mess I had just left, or, indeed, in any other Mess. One or two people I noticed were wearing the ribbon of the DFC. These people I stared at, probably too long, as one stares at celebrities or personalities of importance, hoping to read the signs of some of their experiences written in their faces. But they too looked perfectly ordinary and completely unconscious and oblivious of their distinction. Those talking to them did not seem to be treating them with any particular respect or showing them any deference, but were conversing with them as they might with any ordinary being. I wandered out of the Mess fooling that perhaps life would not be so different, after all.

'I went in search of the duty batman and was told that the Mess was very full at the moment - and that I would have to share a room, I was taken to my room - or, rather, part share of the room - and found the other occupant already in bed and asleep. This other occupant was lying absolutely still and silent and fast asleep. He had scattered his clothes all over the place. Some were on my bed, some were on his bed and some were on the floor. Also on my bed there was an open suitcase, two tennis racquets, a squash racquet and his towel. I removed the articles from my bed to the floor, making as little noise as I could I looked at his tunic, thrown carelessly over the back of a chair, to see if I could gain some clue as to the identity of this unknown person. I saw he was a Pilot Officer like myself; also that he was a pilot. I also noticed that he had no 'gong' up and thought therefore that he, too, might be a new-comer. I could not see much of the sleeper, as only the top of his head, showing brown untidy hair, was visible above the bedclothes. I went to bed, wondering what my new companion and this new life would be like.

'I was awakened next morning by the buzzing sound of an electric razor and saw a slight figure in brightly-coloured pyjamas walking up and down the room trailing a length of electric flex behind him and running the razor in a care-free manner up and down his face. After a few moments I said 'Good morning' and was favoured with some sort of grunt in reply. Undismayed, I started asking questions about the new station and my new squadron but to all my questions the only replies I got were grunts. Eventually I gave up my questioning as a bad job and started to get up. I saw this uncommunicative and, as I thought, strange person several times during the day, but never once did he show that he recognized me. I noticed that he seemed to know everybody and that most people called him 'Cheese'.

'That night I changed my room.'

The occupant of the room that Rivaz was so cautious to avoid waking was

Leonard Cheshire. In fact nothing other than a vigorous shaking would awaken Cheshire once he was asleep. He very rarely went to bed at the average person's time, but either very early or excessively late. At whichever time he went he would sleep until he was awakened and then get up perfectly fresh.

Cheshire would become legendary in RAF Bomber Command, finishing the war with the VC, a bar to his DSO and the DFC. Born on 7 September 1917 in Chester, Geoffrey Leonard Cheshire was the son of Professor Geoffrey Cheshire, a barrister, bursar at Exeter College, Oxford and professor of English law. Leonard was brought up at his parents' home near Oxford and educated at the Dragon School, Oxford and then Stowe School before becoming an under-graduate at Merton College. At Chatham House, Stowe three of Cheshire's closest school friends were later to gain great distinction for individual courage. Jack Randle was awarded a posthumous VC in Assam, Jock Anderson, another posthumous VC while serving with the Argylls and Peter Higgs who served on 111 Squadron RAF, died in action as one of the 'Few' of the Battle of Britain.

Air Chief Marshal Sir Christopher Foxley-Norris was an intimate of Cheshire's for more than half a century. 'In 1936 we were undergraduates together at Oxford, whence he had arrived from Stowe. From neither of these institutions emerged any clear indication of his later extraordinary record of achievement. His Headmaster at Stowe wrote of him 'He is an excellent boy. As a scholar he is hard-working but not very gifted; but he has ambition.' This last word was to prove most significant. At Oxford these ambitions were very much those of the young extrovert hedonist. He coveted high living but did not have the material means to secure it. Perhaps as an approach to attaining it, he sought and attracted publicity by various harmless but irresponsible escapades. At one stage, somewhat ironically in view of his eventual achievements, he avowedly modelled himself on Leslie Charteris's fictional hero 'The Saint', sleek hair, laconic speech, long cigarette holder, Alfa Romeo - the lot.' [15]

At Merton Cheshire graduated with an Honours degree in Law and in 1937 he became interested in flying, joined the university's Air Squadron. Foxley-Norris[16] remembered that 'at first Leonard did no flying. It was at a party in my rooms that I introduced him to my flying instructor, Charles Whitworth. Leonard's reaction was typically casual and sardonic. 'Flying instructor, eh? I suppose you can fly?' The challenge had been laid down and both parties took it up eagerly. Leonard took to flying avidly. He proved to be a natural pilot, but at that stage lack of application prevented him from becoming an outstanding one.'

On completion of pilot training was commissioned in the RAF Volunteer Reserve (RAFVR) on 16 November 1937. He then returned to his studies. He was once bet half a pint of beer by a friend who said he could not reach Paris with just a few pennies in his pocket; Cheshire won his bet. While staying with German friends he witnessed a Hitler rally in Germany, causing great offence to his hosts by not giving the Nazi salute. 'He was one of the comparatively few who not only foresaw the imminence of World War II but did something about it; wishful thinking making ostriches of many of his contemporaries' recalls Foxley-Norris. 'In the summer of 1939 he received what was to be one of the last peace-time permanent commissions awarded by the RAF and as war was declared we both reported to Hullavington for advanced training. But Leonard's adventurism was still alive as was shown when, still not fully trained, he applied unsuccessfully to fight in Finland when volunteers were called for. The training aircraft provided were Harts and Furies for half the course members, the sedate twin-engined Anson for the rest. Leonard drew the Anson, but if he was disappointed he showed no sign of it. Indeed, when on graduation the majority of us opted for fighters or Army Co-operation, Leonard chose bombers. When I asked him, 'What are you up to this time?' he replied, It's pretty obvious isn't it? There's lots of you, there's only one of me. They'll hear about me.' And they did. Within a year he had won the DSO, the first of three and thoroughly deserved it.'

Mobilised on the outbreak of war and after a spell of RAF instruction, Cheshire was awarded his RAF wings on 15 December 1939. He then proceeded as a Flying Officer to 10 OTU, Abingdon and Jurby and on 6 June 1940 had joined 102 Squadron at Driffield as a pilot on Whitleys. His introduction to operations came three nights later when he flew as second pilot to a New Zealander, Pilot Officer P. H. Long DFC over Abbeville. After several more sorties Cheshire soon recognised how little he really knew of the many responsibilities he was about to assume as a bomber captain; a sobering thought which led him to start 'learning his trade' methodically, by consulting his ground and air crews on technical and procedural methods and thoroughly familiarising himself with virtually every facet of his Whitley. For the next five months sortie followed sortie, each adding some aspect of solid experience. [17] 'Interdependence and trust were vital lessons that all operational aircrews had to learn and Cheshire took it further and established a uniquely strong bond of friendship and loyalty with his ground crew also; without it he realised that the dangers of operational flying would be markedly increased. Furthermore he developed a genuine affection for the men who serviced his aircraft; an affection which proved mutual at all levels of age and rank. His aircrew, too, acquired a strong faith in him despite his unorthodox tactics and apparent contempt for danger.' [18]

After breakfast on that first day Pilot Officer Rivaz went up to see the CO 'a charming man who made me feel quite at home and very happy. He told me I was in 'B' Flight and sent me down to see my Flight Commander, whom I later learnt, was familiarly known as 'Teddy'. He was an excitable little man with an enormous backside and proportionately large moustache. As I stood at the door of his office he was on the telephone and I heard him say: 'But I don't want any more gunners: I've got all the gunners I want.' As he hung up the receiver he called me in and said: 'That was you I was talking about.'

'Not so good.' I thought.

'He told me to shut the door and I found him much more pleasant than I first thought he would be as he explained that I should not be put on a crew yet as there were no vacancies and that it was up to me to learn all I could in the meantime. He sent me to see [Flying Officer Edward Denys Stevens] the squadron gunnery leader, who was the oldest and toughest gunner I had as yet seen and who was known as 'Steve'. He was about forty-five and looked as hard as nails, but had two of the kindest eyes imaginable. He was sometimes known as 'Two-gun Steve', as he used to carry a couple of revolvers and a flick-knife which made his tunic stick out from his waist as though it had been starched. Everyone seemed to like and respect him. He had been an observer in the last war; later became a pilot and was now an air-gunner. He had an amazing capacity for work and seemed to expect other people to have the same. He had one of the deepest and loudest voices I have ever heard and was never afraid of using it. He always said exactly what he thought of people and in no uncertain language his vocabulary for swear words being terrific. He took me straight out to an aeroplane to see what I knew, or rather, what I did not know - as he did not seem in the least interested in the little I did know. He spent the rest of the morning teaching me and showing me around.

'I soon got really fond of 'Steve'. If he thought anyone was keen to learn, he would do anything he could to help that person, but, on the other hand, if he thought people were slacking, he would have no further use for them at all. He used to smoke the foulest-smelling cigarettes, which he rolled himself and used to say in defence of numerous protests that when he smoked he wanted something he could taste. He also had one of the largest appetites I have ever seen in anybody. He taught me a tremendous amount about gunnery during my early days with the squadron and used to maintain that everybody should know as much as possible about the aeroplane in which they would have to fly, quite apart from their own particular job. I was not to know 'Steve' for long, as he was killed on an operational trip a few months after I joined the squadron [on 21 November on the trip to Duisburg]. [19]

About a week after my arrival [on 15 August] I was sitting in the ante-room after lunch writing a letter thanking a friend for 'Ming'. Ming had arrived that morning by post and was my mascot; he was a tiny stuffed baby panda and I had him in my pocket while I was writing. The air-raid siren sounded and I looked out of the window and saw people running to the shelters. 'Good lord!' I thought, 'what on earth is all the rush about? The ante-room, which had been crowded a few seconds before, was almost empty and the few remaining were rushing to the door. 'Extraordinary!' I thought. While the siren was still going there came an unearthly screaming noise. All other sounds were then promptly drowned by the largest explosion I had ever heard and the windows of the ante-room were blown in with a din like several rifle-shots. I left my letter and ran to the door - real fun and excitement'. I had never heard a bomb burst at close quarters before and I thought how splendid it was to be seeing some real action.

'I saw someone crouching behind a sofa in the hall as more bombs burst outside and the whole building shook. The next thing I remember was lying on my face in a passage covered with dust and choking and surrounded by broken glass and rubble. I got to my feet and saw through a cloud of smoke that the Mess a few feet behind me was a complete ruin: bricks and plaster, dust and glass, were piled jumbled together. 'This is frightful,' I thought and once more found myself on my face with the roof trembling and shaking as another stick of bombs fell across the Mess - one bursting a few feet in front of me and completely blocking the passage. I lay on the floor gasping for breath; choking and panting while bombs burst all around. I could hear the whine of diving aeroplanes and the scream of falling bombs while all the time the ground shook with the explosions. I was really frightened; more frightened than I had ever been before. I noticed that there was someone else lying on the floor beside me and we clung to each other.

'This is bloody, isn't it?' I said.

Outside there was a new noise added to the din: it was a sort of loud crackling sound and was a building burning just outside. The air was filled with fumes and smoke and dust which were almost suffocating; my lungs felt as if they were dry and empty and I gasped and choked. The inferno seemed to have moved a bit farther off so my companion and I got to our feet and climbed over bricks and stones and rubble and made our way outside. Dust and smoke were everywhere and it was impossible to see more than a few feet. We ran to the nearest shelter and went inside. The first person I saw was the CO, who said to me: 'This is a funny sort of welcome for you!' My tunic was grey with dust and badly torn and blood was trickling down my face. 'You'd better go and see the MO when it quietens down' the CO said again. I remembered Ming in my pocket and decided that that should be his home from then on. He has accompanied me on all my operational trips and still resides there.

When all was quiet I joined the group surveying what remained of our Mess. I noticed that not only had bombs burst within a few yards and on two sides of me, but that one had also burst slap overhead. The roof and wall of the room above where I had been lying had disappeared and I saw a bed standing by itself and clothes strewn all over the debris. Steve joined us, covered from head to foot in dust. He had been lying in a ditch not far from the Mess, watching the fray through a pair of binoculars. He said he had a grand view until he was buried by earth and plaster, which completely obstructed his vision. This infuriated him, particularly as by the time he had extricated himself the show was over!

I went up to the hangars to see what damage had been done there. They had been badly knocked about and one was on fire: the fire party were at work with their hoses amid a great din of crackling and sizzling. In another hangar a Wing Commander was at work with a gang of men all in tin hats, clearing debris from the floor. I went to join them and was promptly bellowed at by the Wing Commander.

'What the hell do you think you're doing? Why haven't you got a tin hat? Can't you see the roof is falling in? Do you want to get brained? If you want something to do, go and help move that Whitley; there's an unexploded bomb beside it! When you've done that, go and wash all that blood off your face and get it seen to!'

I thought that working beside an unexploded bomb that might go off at any moment was far more dangerous than working in the hangar, but I did not say so and went and helped push the Whitley away.

On my way to Sick Quarters I saw a party of men digging furiously around a shelter that had received a direct hit: the ambulance was there and the orderlies were lifting a man - with his tunic, face and hair covered with earth - on to a stretcher. Someone put a cigarette between his lips and lit it for him. Sweat was pouring off his face and caking the earth and I noticed that his legs were in an unnatural twisted position. Someone was digging around another pair of legs: the body was still buried and the legs obviously broken. I saw two more men crushed - with faces nearly the same colour as their tunics - between, sheets of corrugated iron: they were both dead. I decided that my own minor cuts could wait and went to my room and found the windows blown in and a chunk of bomb splinter through my bed!

That night the siren went again just as I was dropping off to sleep. I was in the shelter about a hundred yards away before the siren had finished wailing! I was not the first one there, either!'

Pilot Officer Rivaz on 102 Squadron returned from leave and the first person he saw when he got inside the Mess was Leonard Cheshire. 'I've got you in my crew' he said. 'Grand' I replied. 'Thanks awfully.'

'Don't thank me. You've not flown with me yet,' he said and smiled. His smile was really beautiful: his mouth, instead of getting bigger, seemed to get smaller and his eyes shone. When he smiled he made you feel glad: it was a smile meant for you and you alone.

'We'll have a talk tomorrow' he continued 'and I'll tell you all the things I want you to do. We might be operating tomorrow night.'

I think this was about the first time that Leonard had spoken to me; certainly he had never said as much before. I was really happy. I was in a crew at last and I wanted to tell everyone I saw: I wanted to sing and jump about; to talk to everybody. I felt friendly towards them all.

I was not pleased about being in Leonard's crew particularly anybody's crew would have done; in fact, if I had had the choice, I would probably have chosen anyone but Leonard. But I did not know Leonard then. I don't suppose for a moment that he had chosen me, either: it was just one of those things that happen.

I was going to fly operationally! That was what I kept telling myself and what my heart had been set on ever since I had started to fly. I wanted to fly on operations ... to see bombs burst and see fires!.. to see flak and shoot down fighters! And now I was going to start, maybe to-morrow night! My leave was forgotten and the rather depressed feeling I had had when I entered the Mess: everything was forgotten except that I was going to fly over Germany!

When Leonard said 'I've got you in my crew' I became so excited that I stayed awake a long time that night thinking about my new fortune and wondering where I should be at the same time the following night.

I sat next to Leonard the following morning at breakfast and found him already midway through a huge bowl of porridge. He always seemed to have a big appetite and had an amazing way with waiters and usually, in consequence, got far more to eat than other people. I often heard him say in his quiet, serious voice to one of the Mess waiters: 'I don't think I can manage three eggs this morning, two will be enough!' and he will be brought two while the rest of us are given one! He has his stock phrases of humour which he never tired of using; I had heard them dozens of times and I hoped I would hear them dozens more times! Most people never knew if he was being serious or not: he spoke in such a serious voice and with such a serious expression on his face. One of his pet remarks was - when he was about to sit down next to someone already seated - 'Don't get up!'- at the same time raising his hand as though that person was about to rise. Another time, when you were reading a letter; he would lean across and say to you: 'After you with that when you've finished.' It rather nonplussed him if you passed the letter across to him! Or again, if he heard someone say 'Good God!' he would say: 'Yes, what is it?' The person would probably say - or so Leonard hoped - He only said 'Good God'.' Leonard would then reply: 'Oh I thought you were addressing me!' All these expressions had been tried on me many times and I have now got the right answers!

'We are on tonight,' Leonard said, turning to me.

'Splendid! D'you know where?'

'Yes, but keep it quiet.' Italy!'

'Oh, boy what a magnificent trip to start off with!

Leonard had plenty of work for me to do that day: he wanted to know how much I knew about the aeroplane. How much Morse did I know? Could I find everything in my turret in the dark? Did I know how to launch the flares? What -did I know about dinghy drill?' Did I know how to bail out? Could I map-read? Could I operate the emergency hand-pump for the undercarriage? Did I know where the petrol cocks were?, and so on.

He introduced me to the rest of his crew. There was Desmond Coutts, our second pilot, a large fair-haired man of about twenty-two. He was magnificently built and seemed to be in a permanent daydream and I thought he looked more like a poet than a pilot; I had often seen him before, exercising a bloodhound around the aerodrome. Then there was Taffy, our navigator, all smiles and friendly at once. The other member was Stokie our wireless operator and, as Leonard said, one of the best wireless operators in the Air Force and afraid of nothing. Leonard was very proud of his crew and I hoped that one day I should come up to the standard of the others.

We did our NFT that morning and Leonard kept calling me up on the intercom. I was beginning to feel that I was very fortunate in my new pilot. After we landed he wanted to know what I thought of the turret and if everything was all right. All that day I kept saying to myself, 'We're going to Italy tonight!'I could not have been more excited if I had been, going there on a holiday, or if I had been going to some marvellous party. Life, I felt, was just beginning: I was a member of an operational crew and would soon be doing my first operations! This was what I had been waiting for and for what I had been longing!

It started to rain that afternoon and was still drizzling as we dressed in our flying clothing: Leonard asked me if I had on plenty of clothes, as I would probably be cold. We drove out to our aeroplane in the drizzling rain and had to wade the last hundred yards or so through mud. The ground crew were waiting for us, sheltering under the wings. We quickly got inside out of the rain and I went straight to my turret. Almost as soon as I had settled myself a message was brought saying that the operation had been scrubbed. I will not attempt to describe my disappointment nor the anger of the others.

The following night, 26/27 September, we were briefed to go to Dortmund in the Ruhr. I was disappointed it was not to be Italy but, as Leonard said, there would be plenty of other opportunities; anyway, I was delighted at the thought of going anywhere. All that day I kept looking at the weather and was on tenterhooks lest we had a repetition of the previous night. However, we did get off all right this time. It was dark when we took off and I thought, very mysterious and exciting: I had flown many times at night before, but never on such a marvellous mission. Until we were actually airborne I was worried lest the trip should be cancelled at the last moment. I was interested and thrilled with every word spoken down the intercom. It was all important and seemed to presage great things to come.

It was a pitch-dark night and I could not see anything on the ground below me. Leonard and Taffy had quite a lot to say about the navigation: I kept silent, absorbing it all, rather as a child takes in the conversation of its elders. Occasionally Leonard called me up to ask how I was or if I was cold; after we crossed our coast he told me to test my guns. I had them all ready loaded and cocked and pressed the trigger. There were yellow flashes from the muzzles of the guns as they fired and I watched the incendiary bullets - looking like illuminated dots - leaving the guns and retreating rapidly in a curve as they disappeared into the night. There was a smell of burning cordite which I knew quite well and rather liked; it overcame for a few minutes the rather sickly oily smell peculiar to gun turrets. I told Leonard that all the guns were OK and he replied 'Good show!'

Desmond flew most of the way over the sea, but Leonard took over again as we crossed the Dutch coast. I could just see the coast line below me and felt a peculiar thrill as I realized we were over enemy occupied territory. Leonard told me to let him know if I saw any lights or anything. Everything from now on would be completely new to me and I was very excited and ready to be entertained!

'I saw some searchlights in the distance which looked exactly like our own, except that in amongst their beams there were little flashes of light rather like jewels sparkling and I thought rather pretty. I told Leonard what I saw and he said we should probably see some a lot closer later on. I saw several lights on the ground below us, all of which I reported,: some were quite bright lights, while others I could hardly see but they all interested me intensely and I tried to imagine what they were and what was happening below. When I was not glancing at the ground I was staring out into the night, wondering if I should see any fighters. I kept looking about me up and down to the front and to either side: the moon had risen but was in its last quarter and I wondered how far I really could see. Until one is thoroughly used to searching the sky at night one's eyes get very dazed and tired, probably because one is trying to take in too much but with practice it is possible to focus the eyes at a given range and look, as it were, at something instead of merely staring out into the distance, when one can be looking yet not really seeing. Try sitting in a dark room and looking out into the night for eight or nine hours and you will see what I mean.

As we neared the target Leonard and Taffy had further discussions as landmarks they recognized appeared or landmarks they expected to see did not appear. I tried to make something out of those light and dark patches on the ground which seemed to mean so much to them. Taffy said our ETA was up and we should be over Dortmund. Leonard started to fly round in a wide circle and cursed the darkness. We continued to fly round in a circle for some time, with occasionally Leonard or Taffy making some remark. I thought it seemed very tame and quiet, as I had imagined we should be seeing searchlights and flak and all sorts of things and I was rather disappointed.

Leonard must have spotted the target, for he said. 'There it is; down on the port side. I'm going to flyaway and come in right over it! Get ready to drop the bombs, Taffy.'

I felt a renewal of excitement and wondered if I should see the bombs burst: I wanted to very much. I could hear Taffy giving Leonard directions over the intercom as he watched the target through the bomb sight. At each turn that Taffy gave I could feel the aeroplane instantly change from its course.' Taffy was saying 'Left...left...steady! Right! ... Steady ... steady ... Bombs gone!'

Almost at the same moment that Taffy said 'Bombs gone!' I saw a number of bright flashes on the ground: there seemed to be hundreds of them ... and almost immediately we were surrounded by flashes and crashes and bangs and seemed to be the centre of some wild excitement of lights and noise. Leonard must have been doing some violent things with the aeroplane, for the ground - now clearly visible by its flashes of light - instead of appearing below us, seemed to be tilting from side to side and at times appeared to be almost overhead. Now and again the aeroplane lurched violently and then dropped, as an explosion louder than the rest was felt and heard below us. So this was what flak was like! I began to wonder if it was much fun, after all in fact, I thought it looked and sounded extremely dangerous.

'I've only dropped one stick' I heard Taffy say.

'OK' Leonard replied. 'I'll fly in again. Let's see if we can get rid of this stuff first, though.'

'They don't seem much concerned,' I thought. 'Perhaps to them this is just an ordinary occurrence; something they experience each time they bomb.' I did not know if it was more violent than usual or average or perhaps even rather mild: what I did know was that I did not like it! I had heard stories of flak. 'Almost blew us out of the sky!' 'Could feel it hitting us!' or 'Came back full of holes!' But these words meant nothing to me. I had seen aeroplanes come back with holes through them and rather envied the crews their experiences but I had no idea what it would really be like and now I was beginning to find out. There were dozens of searchlights on us and every dive or climb or turn we did we were still held in their beams. Hundreds of what looked like red and yellow and green balls of fire were hurtling past us and around us: sometimes they were so close I felt I could stretch out my hand and catch them. I began to feel that each one was aimed at me and would burst through my turret but as each one seemed on the point of hitting me it somehow miraculously skimmed past just clear of the tail.

An aeroplane from the ground always looks so small but when you were flying in one and being shot at, it always seems so large. I was fascinated by this cascade of destructive little lights coming up at us so relentlessly all the time: they seemed to start their journey upward so slowly but as they came near they whizzed by with incredible speed. I could see the flashes from where those guns started and fired my guns at them as well as at the searchlights. I fired hundreds of rounds and, for a few minutes, went mad, shouting and cursing and laughing all at once. What was I there for? Why should I not shoot, too? Why should I sit still and be shot at when I had some guns in front of me? I hoped I was hurting them ... killing them ... making them run! While all these hundreds of coloured balls of fire were popping up at us, larger shells - invisible until they burst-were hurtling through the night. They burst with blinding flashes and with dull, muffled thuds that shook the aeroplane and hurled me hard up against the side or top of my turret.

'I'm going to turn round and run over the target again.' I heard Leonard say to Taffy. Still the shells stayed with us and seemed to be, if anything, increasing. I began to sympathize with the pheasant who flies down the line of guns, being fired at by each gun and still continues on apparently unscathed. But I knew how many are not really unhurt but drop dead or mortally wounded some distance farther and I wondered if we were like that and would do the same.

I began to feel rather lonely and cut off from the others and wondered what was happening at the other end of the aeroplane. Perhaps even now they were clipping on their parachutes, preparing to jump; perhaps they had already jumped and forgotten all about me.

Almost as if he had been reading my thoughts Leonard said, 'How are you, Revs?'

From that moment on, Leonard has always called me 'Revs'.

Why, I have no idea, as most other people call me 'Riv'.

I managed to croak back some sort of reply. Those four words bucked me up no end. I felt again that I was not the only person in the aeroplane and once more felt part of the crew. I wondered how the others were feeling and if they were taking it as a matter of course.

I was still firing bursts at the guns and searchlights and had a look at my ammunition. It would not do to use all my ammunition here; I should have to leave some for possible attacks by fighters on the way back that is, if there was to be any on the way back. I wondered how we had got off free for so long and felt the next one would surely bring us down.

Once again I heard Taffy give his lefts and rights. I wished he would hurry up and drop the bombs so that we could get clear away from all this. I had no idea how long we had been diving and twisting about in this fountain of fire and steel, but it seemed like all night. It seemed we had always been doing this and would always continue to do it.

Once more I heard Taffy say 'Bombs gone' and thought 'Thank God for that!' I looked hard, waiting and hoping to see them burst. Suddenly, in amongst the flashes, I saw what looked like a sheet of flame and in the midst of the flame, blue and white flashes like lightning. Those flashes continued for several seconds and completely outclassed the gun flashes.

At last we were clear of what Hell must be like and were flying normally once more and I heard Leonard say, 'Someone bring me some coffee.' I felt I could do with some, too my throat and mouth were dry and my turret was full of the fumes of burnt cordite from all the rounds I had fired. I also noticed that my clothes were sticking to me with sweat. Once more we were flying undisturbed through the night and once more I realized there was a moon and felt a sense of tremendous relief.

I heard a banging against my turret doors and put my hand behind my back to open them. There was Stokie with a flask of coffee and grinning his hardest.

'We've had enough for one night,' he shouted in my ear and I hoped he was right. I saw the Dutch coast on our return with very different feelings from those I had had on the way in and was not at all sorry to see it disappear into the darkness.

It was just getting light when we landed and examined our plane from the outside. The wings, tail and fuselage were full of holes and we counted over a hundred of them.

I went to bed after my first 'operational breakfast' but did not stay there for long as I found I could not sleep: flashes and bangs accompanied me whenever I started to doze off. Evidently Leonard had not been able to sleep either, for I found him in the ante-room writing letters. That afternoon we went to York to see *Top Hat*.

Two nights later, on 28/29 September, we went right down to the south of Germany, to Leuna and were flying for nearly eleven hours. The next trip was to Düsseldorf and then Essen, Duisburg and so on... Then came the full moon period and Cologne on the night of 12/13 November. I far preferred operating when there was a moon: you don't get that enclosed and rather oppressed feeling that you get on dark nights when for the majority of the time you can see nothing but stars and sometimes not even them. We had a new wireless operator [Sergeant 'Taffy' Roberts] in place of Stokie, who had been posted.

When we crossed the Dutch coast I could see every detail, with the sea looking like silver; stopping suddenly as it reached the mud flats and then dividing up into hundreds of little glistening streaks as the moon glanced on the dykes and drains. Sometimes I saw what looked like lakes but were, I imagine, floods. We saw the Rhine perfectly, winding and curling about and Leonard followed it for some time: it looked so serene and quiet in contrast to all the flak being pumped up at us. I thought what a rum game it all was, all the hundreds of shells being hurled up at us while we were waiting to send a load of bombs hurtling down. Taffy must have been having some trouble with his intercom or something for Leonard kept asking him where we were, but could get no reply.

The flak suddenly closed down and I heard Leonard say something about a trap; probably fighters. I was watching out very hard, but could see no sign of any fighters. It was brilliantly clear and I should certainly have seen them if there had been any near. Taffy must have done something about his intercom for I heard him talking to Leonard. We were too far north, so turned south. All the time I looked about me ready for fighters which I did not see. It was certainly very quiet; unnaturally so: we must be practically over Cologne but not a gun: only ourselves and the brilliant moon. The moon was to one side of us and I could see it by leaning forward and turning round: I wondered how far I could really see certainly several miles. Leonard was cursing some low cloud and I looked down and saw it like white fleece below us.

Suddenly the world seemed to stop or else race ahead! I remembered a deafening explosion and a blinding red flash which seemed to be inside my head and behind my eyes. I was falling through darkness falling to the ground. I felt I was still in my turret, but could not .see it; everything was dark and silent and the engines had gone. I knew we had been hit and I imagined the shell must have burst somewhere behind my turret and blown it from the aeroplane and I was falling in it! So this was the end! But it was not the end yet, as I was still falling. I wished it would hurry up and finish; it took so long. I had been falling for such a long time and everything was so quiet and I was all alone. Should I feel when I hit the ground? I must be nearly down by now, surely! If only I could see! But no, I would rather not see. I would rather it ended quickly and not know about it.

'Revs! Revs! Come forward Revs!' So Leonard was here, too! But how did he get here? He was in the front! Come forward, Revs.' What did he mean, come forward? There was the moon but it was red instead of silver. What were all those coloured lights? Hell! They were shooting at us. Flak; always bloody flak! Why hadn't I hit the ground yet? But I was not falling anymore and Leonard had spoken to me: he said 'Come forward, Revs.' I could not leave my turret unless he told me to but he had told me to and I wanted to leave very badly: I was frightened and wanted to see someone and feel someone.

We were still flying and the engines were running. I put out my hands and felt for the sides of my turret. It was still there - yes - and whole. I must go and see what had happened. The door behind me was shattered to splinters and there were smoke and flames, too. I must do something quickly. Leonard had called me forward and he wanted me. I crawled forward through the smoke and flames.'

God, what a mess! The fuselage door had gone and most of one side of the fuselage as well. Desmond was there, working like a maniac, with his blond hair shining in the light of the flames and his eyes sparkling like brilliants: sweat was pouring off his face and he was hurling flares, incendiaries and spare ammunition out of the gaping fuselage. I started to do the same and he shouted at me to go back and get my parachute, as the aeroplane would probably break in two at any moment. 'Very probably,' I thought and wondered how it was holding together even now. I hoped it would not give way before I had, time to crawl back for my parachute. Somehow the idea of jumping did not worry me it bit; it seemed inevitable that we should have to do that and it would be so simple just to fall out over the side. I hardly noticed the flak and was only dimly aware of the crumps and flashes all around us.

When I got back to the fuselage with my parachute the flames had nearly all disappeared, but the place was still dense with smoke. I worked with Desmond until the flames had completely gone and then went forward. The wind and slipstream was whistling through the fuselage and tore at my clothing as I crawled along. Leonard was sitting at the controls and turned round and smiled as I entered the cabin. Davy [Sergeant Henry Davidson] was sitting by his set fumbling for the Morse key: his face was charred and black and his clothing all burned. I looked at his parachute harness and examined the straps. If he had to jump I thought they would probably hold, as the flames had done little more than scorch them, although the surrounding material was nearly all burnt away. Davy looked very bad: he did not look as though he could pull the ripcord on his own and I started to think out plans for getting him out safely.

It was almost as draughty in the cabin as it was in the fuselage and I tried to get Davy to lie down on the floor, where it was slightly warmer but he would not leave his seat. Once I forced him to lie on the floor but all the time he struggled to get back into his seat and at last I had to give way to him. His one idea was that he must carry on with his job and not let us down. He had been standing in the fuselage by the flare-chute, ready to drop a flare, when the shells burst. A splinter from one touched off one of the experimental flares we were carrying; it exploded with a flash of several million candle-power and how it did not blow him to bits or hurl him out of the fuselage - which was split and left with an enormous gaping hole down one side I cannot imagine.

Poor Davy was very badly burned - chiefly about the face - and was quite blind. Desmond had fetched the first-aid kit and covered Davy's face with the jelly used for burns. It was bitterly cold in the cabin and for the next five hours we did all we could for Davy's comfort: I kept putting his fingers in my mouth and breathing hard on them to try and get them warm. He kept asking me where we were. When we were still over Germany I told him we had crossed the coast and were over the sea and when we were over the sea I told him we were over England and would soon be back. He must have thought the last part of the journey took an interminable time!

It was a nightmare journey, with the five of us crammed into the cabin of the Whitley, hardly ever speaking and wondering how far we should get. Leonard was sitting at the controls; he had taken his helmet off, but was still wearing his yellow skull cap, which looked grotesque in the half-light. Taffy was sitting at his navigation table grinning to himself most of the time: Taffy always grinned under all conditions! Desmond, looking like a wild blond giant, was part of the time sitting beside Leonard and part of the time looking at Davy. Davy sat at his set, hardly moving; I was crouched in the confined place beside him. As I watched his face I tried to remember what he had looked like before but his charred black mask gave me no help.

The journey back seemed endless, but, like everything else, it did end. We crossed our coast in broad daylight and after landing were surrounded by an inquiring crowd. I did not go to bed, but went into York with Leonard and Desmond. We went to a cinema again in the afternoon and I slept through most of the show.'

A few days later I was sent on a Turret Course, which lasted a week and when I got back Leonard [who was awarded an immediate DSO for bringing his Whitley home from Cologne while Davidson received a DFM] told me he would shortly be leaving the squadron and going to a Halifax squadron. Naturally, I wanted to go with him and he said he would try and fix it. I also did all I could to get posted with Leonard, but without any luck at all.[20] When I finally knew for certain that I should not be going with him, I set about to get myself fixed with a new crew. Desmond Coutts had been given a crew, so I asked him if he would have me as his gunner. I think he was genuinely pleased with the idea, but he said he was fixed for a gunner for his first trip - which was that night - but would take me next time. There was not a next time as Desmond was killed that night. There were no more crew vacancies: and I had a spell with practically no flying. I was made Flight Gunnery Leader, which meant a certain amount of work on the ground - chiefly organizing. I wanted to get back into the air and was impatient with this spell of inactivity.'[21]

Endnotes Chapter One

2   *The Burning of Frankfurt: A German City and Its People 1939-1945* by Rafael A. Zagovec.

3   Wing Commander Bintley DSO AFC was killed on the night of 23/24 October 1942 at the controls of his Halifax on 102 Squadron when he was in a collision with another 102 Squadron Halifax at Pocklington on the operation on Genoa. His Halifax crashed at Barnby Moor killing Bintley and Flying Officer A. J. Graham. Two other crew were injured.

4   Flying Officer Cogman and his crew were shot down on 19/20 May 1940 on the operation on Gelsenkirchen. The Whitley crashed at De Klef ENE of Eindhoven. Cogman evaded and three of his crew were captured. One man was killed. Raphael and his crew ditched 60 miles off the Dutch coast on 18/19 May when they were hit by a Bf 110 returning from the raid on Hannover. Raphael, who had a painful foot wound and the crew, all of whom were injured, were rescued after four hours afloat and taken to Great Yarmouth. *RAF Bomber Command Losses of the Second World War, Vol.1* by W R Chorley (Midland).

5   Budden was killed on 11/12 June 1940 on the operation on Turin. Flying Officer Gould was killed on 5 April 1940 during training when his aircraft crashed in the Bristol Channel.

6   Sergeant Ernest Short MiD and Flight Sergeant Albert James Heller DFM were crewmembers on Pilot Officer Ernest Ronald Peter Shackle Cooper's Halifax crew on 35 Squadron on the night of 25/26 July 1941 when the aircraft was shot down on the operation on Berlin and they were all killed.

7   Adapted from the 51 Squadron official history in *The Right of the Line: The Royal Air Force in the European War 1939-1945* by John Terraine. Hodder & Stoughton 1985) and B*omber Command; The Air Ministry Account of Bomber Command's Offensive Against the Axis September 1939-July 1941.* (HMSO 1941) Warrant Officer Tom Bowles MiD was shot down on the operation on Jülich on the night of 21/22May 1940 and was taken prisoner with his crew. He died of tuberculosis in captivity on 26 April 1941. Flight Lieutenant Emery DFM was KIA on the night of 4/5 April 1943 piloting a 51 Squadron Halifax on the operation on Kiel.

8   Adapted from *The Paper Raiders* by Peter Williams writing in *RAF Flying Review,* January 1961.

9   *Feuersturrn uber Hamburg* by Hans Brunswig, published by Motorbuch Verlag, Stuttgart.

10   *The Burning of Frankfurt: A German City and Its People 1939-1945* by Rafael A. Zagovec.

11   Quoted in *Experiences of War: The British Airman* by Roger Freeman (Arms & Armour 1989).

12   Kinloss proved to be the first from which crews were sent to squadrons in 5 Group, which were re-equipping with Lancasters. From Kinloss, Shannon was posted to 106 Squadron where Wing Commander Guy Gibson had recently been appointed to command. In May 1943 both men flew the Dam's Raid. Quoted in *Out of the Blue: The Role of Luck in Air Warfare 1917-1966* by Laddie Lucas (Hutchinson 1985).

13   Oxley was soon promoted to Wing Commander and he took command of 50 Squadron.

14   Hayhurst was awarded the DFM. Connor was killed on 3/4 November 1940 when his Hampden crashed off Spurn Head returning from a raid on Kiel. His three crew died with him.

15 *Leonard Cheshire VC and OM: The Unanswerable Double* by ACM Marshal Sir Christopher Foxley-Norris, quoted in *Thanks For the Memory: Unforgettable Characters In Air Warfare 1939-45* by Laddie Lucas (Stanley Paul 1989). Cheshire is the only man ever to hold the Victoria Cross and the Order of Merit - the unanswerable double.

16 After Winchester and Trinity, Oxford, a Harmsworth Scholarship to the Middle Temple and an intended career at the Bar frustrated by war Sir Christopher flew Lysanders in the Battle of France in 1940, followed at once by fighters in the Battle of Britain and then, later, lethal anti-shipping strikes on Beaufighters and Mosquitoes against naval and other maritime targets in the Skagerrak and Kattegat and along Germany's north-western seaboard.

17 *Bomber Barons* by Chaz Bowyer (William Kimber 1983).

18 *Leonard Cheshire VC and OM: The Unanswerable Double* by ACM Marshal Sir Christopher Foxley-Norris, quoted in *Thanks For the Memory: Unforgettable Characters In Air Warfare 1939-45* by Laddie Lucas (Stanley Paul 1989).

19 'Steve' Stevens was flying as tail gunner on Whitley V P5072 flown by Flying Officer Anthony Marmaduke Langdale Selby which crashed in the North Sea with the loss of all five crew which included Wing Commander Samuel Robert Groom.

20 Cheshire was genuinely indignant that only he and his wireless operator received decorations; unfortunately the award of his own DSO revived some of his old hunger for publicity. By January 1941 Cheshire had completed his first tour of operations, but immediately volunteered for a second tour and was posted to 35 Squadron at Linton-on-Ouse. The first unit to operate the Handley Page Halifax, 35 Squadron had reformed in November 1940 and flew its first Halifax sorties in March 1941. That same month Cheshire was awarded a DFC and in April he was promoted to Flight Lieutenant. From April to July 1941 he was detached from the squadron for ferrying duties to and from Canada. On leave in New York Cheshire met and married 41-year old Constance Binney, a wealthy divorcee and sophisticated socialite once America's top silent movie star and successor to Mary Pickford in 1922. Contrary to general predictions, they both derived real if short-lived happiness from the marriage. Cheshire then resumed operations on 35 Squadron, being further promoted to Squadron Leader in October 1941 and completing his second tour of 'ops' by early 1942. His 'rest' consisted of a posting to 1652 HCU at Marston Moor as an instructor, though during the following four months he managed to fly four more operational sorties, including the first of the famous 1,000-bomber' raids, on Cologne. By August 1942 Cheshire had returned to full operational flying, with 76 Squadron, another Halifax bomber unit and was appointed commanding officer with the rank of Wing Commander. By then he had been awarded a Bar to his DSO. He remained as CO of 76 until finally leaving the squadron in March 1943 on promotion to Group Captain - at the age of 25 and becoming the youngest group captain in the RAF. By then he liked a suite at the Ritz on leave and to bask in The Mayfair cocktail bar. He took up an appointment as commander of Marston Moor RAF station. When he was offered an opportunity to return to operations in September 1943 - albeit with reversion to Wing Commander rank - he accepted eagerly. His new post was to be as commander of 617 Squadron at Woodhall Spa and tasked with 'specialist' operations only by higher command. On taking up his command, however, he was privately disturbed by higher authority's insistence that he train 617's crews in high level bombing technique. Throughout his long operational career Cheshire had been a firm advocate of low-level attacks, thereby improving accuracy. Cheshire seized the first real opportunity to demonstrate his belief in low-level attacks on the night of 8/9 February 1944 when Cheshire led a dozen of his Lancasters to bomb the Gnome & Rhône aero-engine factory at Limoges 200 miles southwest of Paris. The factory was undefended except for two machine guns and Cheshire made three low-level runs in bright moonlight to warn the 300 French girls working the night shift to escape. On the fourth run he dropped a load of 30lb incendiaries from between 50 and 100 feet. Each of the other eleven Lancasters then dropped a 12,000lb bomb with great accuracy. Ten of the bombs hit the factory and an eleventh fell in the river alongside.

While no award of the Victoria Cross was ever made for a Mosquito sortie, Cheshire's contribution to the success of the Munich operation on 24/25 April 1944, when he led four Mosquitoes of the Marking Force in 5 Group, was mentioned in his VC citation on 8 September 1944. In part it said, 'Cheshire's cold and calculated acceptance of risks is exemplified by his conduct in an attack on Munich in April 1944. This was an experimental attack to test out the new method of target marking at low level against a heavily defended target situated deep in enemy territory. He was obliged to follow, in bad weather, a direct route which took him over the defences of Augsburg and thereafter he was continuously under fire. As he reached the target, flares were being released by our high-flying aircraft and he was illuminated from above and below. All guns within range opened fire on him. Diving [from 12,000 to 3,000 feet and then flying repeatedly over the city at little more than 700 feet] he dropped his markers with great precision and began to climb away. So blinding were the searchlights that he almost lost control. He then flew over the city at 1,000 feet to assess the accuracy of his work and direct other aircraft. His own was badly hit by shell fragments but he continued to fly over the target area until he was satisfied that he had done all in his power to ensure success… for a full 12 minutes after leaving the target area he was under withering fire, but he came safely though… What he did in the Munich operation was typical of the careful planning, brilliant execution and contempt for danger which has established for Wing Commander Cheshire a reputation second to none in Bomber Command'. Cheshire flew his 100th op on 8 July and he was withdrawn from operations.

21   Flying Officer Desmond Christian Frederick Coutts and crew were killed on the night of 2/3 January 1941 when their Whitley crashed 30 miles east of Withernsea, Yorkshire on the operation on Bremen.

Chapter 2

Blitz Nights

'It was a shocking experience, but people just got on with their lives. I never saw any loss of morale. I saw a lot of people determined like I was - I had a friend who was a little Welshman and I said to him one day, 'You didn't need to be here, you could have gone back to Wales.' He said, 'No. In the depression Coventry gave me a job and I am not leaving!' And that was the spirit of the people.

R. J. Paxton, an engineer who serviced and repaired machines used to produce Merlin aircraft engines. He also served as a Fire and Rescue Volunteer during the Coventry Blitz.

In 1931 Mr Neil Bell, in a book with the title *Valiant Clay*, predicted the outbreak of a world war on Sunday, 3 September 1939. 'In the dark streets the burned and the wounded, bewildered and panic-stricken, fought and struggled like beasts, scrambling over the dead and dying alike, until they fell and were in turn trodden underfoot by the ever-increasing multitudes about them... In a dozen parts of London, that night, people died in their homes with the familiar walls crashing about them in flames; thousands rushed into the streets to be met by blasts of flame and explosion and blown to rags; they came pouring out of suddenly darkened theatres, picture-houses, public-houses, concert and dance halls into the dark, congested streets to be crushed or burnt or trodden to death, or to meet a swifter end from bomb, or the rain of broken masonry that killed and buried them at one stroke. With the failure of the light the electric power failed also and packed trains in the underground tunnels waited in long-drawn-out suspense that gave place to panic and horror as here and there in weaker places the roofs collapsed from the impact of bombs overhead. And presently fire came to add a last touch of hellishness, to that mad struggle, in which frenzied men, women and children wedged themselves into ever thickening and deepening masses that at last were motionless and silent except for hands that waved and clutched feebly and intermittent choking cries... Next morning the first hospital train loaded with wounded was steaming out of Liverpool Street when the first gas bombs fell on London and the second German attack had begun...The bombs were gas, HE and incendiary and before the punishment ended, London, with its environs and suburbs had become a place of ruin and sepulchre so vast that in comparison, Sodom and Gomorrah, Herculaneum and Pompeii, were but ant-heaps scattered by the feet of children.'

Throughout the summer of 1940 while Bomber Command kept up its attacks on especially vulnerable or especially important parts of the German economic system, Luftwaffe raids on London were at their height. The East End of London, stretching along the Thames for six miles and inland for a depth of three miles was one of the worst slum areas in the western world. Between the wars various schemes had been undertaken to improve things, but the root cause, poverty, had not been addressed. So no headway was made. However, the 'phoney war' was now over and the Blitz was about to demolish many of the slums. Many of those who were made homeless never did come back to the East End. Annette June Coppard, then an East London schoolgirl, had witnessed aerial activity in the skies over Biggin Hill during the Battle of Britain as Hitler stepped up his campaign to annihilate RAF Fighter Command as a necessary prelude to his planned invasion of Britain. Annette's father was stationed at Kenley airfield, not too far away. He worked as an aircraft mechanic. As the conflict intensified, ground crews and pilots alike were perpetually exhausted. Sometimes the pilots were so worn out; they fell asleep as they ate. The skies seemed always to be filled with planes and the noise was tremendous. It was a different world from the peaceful skies of Sussex she had known. In September Annette and her sister Clare started back to school at Hawes Down junior. If the air raid siren sounded while school was in session, the children picked up their books and marched down the steps into the underground shelter to continue their work. The light was dim and when winter came it was bitterly cold with water running down the bare brick walls. A dank smell permeated everything. Air raids were frequent. One day they went to the shelter three times. If there was an air raid at night and the all clear sounded before midnight then school was open as usual next day. If the raid occurred after midnight, school was closed. Several times they walked the mile to school, only to find it closed. Once the school was closed for two days because of an unexploded bomb in a nearby street. 'It was all very stressful and confusing' recalled Annette. 'As the assaults on London escalated, we ceased going to school very much at all.'

'On 7 September hundreds of bombers targeted the London docks, setting them ablaze, lighting the way for more destruction throughout the night. We lay listening to wave after wave of bombers passing overhead. The racket of the anti-aircraft barrage was tremendous, but we found it comforting. The noise gave us the impression that something, at least, was being done to deter the bombers. Because of the din we couldn't hear the noise of the bombs whistling down, but neither could we sleep. The morning after a raid, Clare and I would rush outside to pick up pieces of shrapnel. We vied with each other to see who had the more interesting shapes of metal in our collections. On the afternoon of 15 September, when every aircraft that RAF Fighter Command could muster was flown to fight off the Luftwaffe attacks on London, the Battle of Britain was on its way to being won. Two days later, Hitler postponed the invasion of Britain. It never did take place.

'The Blitz began in earnest that autumn. For fifty-six nights in a row London was bombarded from dusk to dawn. In our dining room we had a Morrison shelter. It was like a steel table and we slept underneath on a mattress. After a while we didn't even bother to go to our beds upstairs because we'd always have to come down to the shelter when the siren sounded. Night after night Clare and I lay either side of Jennifer as she cried in fright at the noise. It was okay to cry when you were only three. Clare and I were frightened, too, but we took comfort in the fact that we were all together. Another 'safe' sleeping area was made in the big cupboard under the stairs, which Granddad reinforced, with solid wooden beams. An additional sleeping place was under the heavy oak dining room table. Along with the Morrison these places would have provided protection from falling debris had our house been damaged.'

Margaret Holmes, a vicar's wife living on the Isle of Dogs in east London, vividly recorded the effects of the Blitz in a letter she wrote after the first onslaught of bombings.

'All last week we had dozens of warnings - six or seven in a day and often lasting all night. Most people around us were frightened and went to shelters for every 'raid', taking their beds there at night. The planes only seemed to be reconnaissance flights and no bombs were ever dropped. Then last Saturday evening at 6 pm, Arthur [the Vicar] was out visiting (the warning had gone an hour ago) and I was watching from our front door for any signs of activity, when I saw a formation of planes coming over. They were mostly huge black ones with four engines. I'd never seen any of this kind before and they were much lower than usual. Our guns battered them, but they kept beautiful formation until they seemed to be overhead, when we went home we retired to our bed under the stairs. It was an awful night, lying down under that stair, listening to bombs near and far and wondering when one would come for us! We soon got to know the whistle and crash of the high explosive, the long whistle and awful shaking (but no crash) of delayed-action bombs. Then at 3.15 am we thought ours had come. The crack was terrific, the front door burst open, broken glass crashed on the pavement and dust and stones rained down everywhere. We looked out and could see nothing but dust, so we waited. I went through to Dodger, who was in his chair in the scullery. The back door had burst open too, but Dodger shook himself and wagged his tail. The dust settled and we looked out. The street was carpeted thick with dust, bricks and glass. Arthur ran round the corner and met Fire-master Ayre, who said, 'Was anybody in the church?' Arthur said, 'No - why? Has it gone?' 'Yes.' Arthur said, 'Oh dear. Unemployed.'

'Then Arthur took me round to see the church. What a sight! A crater, I think about 15 feet deep, took up the width of the road. It was floodlit by a 12-inch gas main burning nicely; the water was pouring in from the water main. All round the crater was spread half the church. The altar end was left standing and you could look up to the altar. Windows and ceilings were broken for hundreds of yards along the street, where pieces of church had flown about. The pub at the next corner had all the ceiling down and all glass out and yet the landlord was selling beer at 5 am.'

London got a temporary respite when the Luftwaffe began attacks on other areas of England. The lull continued over Christmas. Then on 27 December, the sirens sounded again and once more fire bombs began to drop over the

City. It was a fairly bad night. Londoners thought that they were in for it again. But the following night nothing happened. The sirens were silent. The man in the street breathed his sigh of relief and speculated on the chances of a resumption of the lull. But twenty-four hours later these speculations were rudely put to an end. On Sunday 29th December the Luftwaffe fired the age-old City of London in the most savage attack of the aerial war. The German communiqué said that 100,000 fire bombs were dropped. The air-raid warning was received at approximately 6 pm in the control room at Fire Service Headquarters. Soon after, the City of London report centre telephoned advice of two large falls of incendiary bombs in the E.C. district and to the north and south of the Guildhall. In a very few seconds further reports of incendiary showers were received from other parts of London. Soon local stations in these areas were inundated with fire calls and emptied of their first-line engines. About an hour later a serious fire situation had developed in the neighbourhood of St. Paul's Cathedral. Towards 8 pm two further conflagrations were reported spreading. Fire was spreading easily in the City danger zone where the buildings were old and particularly open to fire risk, where narrow alleys and crooked streets ran between warehouses crowded with inflammable stocks, where space was so valuable that courtyards were roofed over with glass to house more and ever more sacks and crates packed with easily-fired goods. An adequate organisation of roof spotters would have saved many buildings and much stock from the peril of sparkstorms. As it was, there were few roof spotters and the fire spread. In addition to this, the owners of many buildings had padlocked and bolted their doors, thus seriously hampering the firemen.

There was normally no shortage of water for the fire services in London, but on this occasion immediate calls were sent for the supply of emergency water. A time-lag necessarily occurs before this water can come through. Pumps must be positioned on the Thames, dockside, canals, lines of hose laid to the fire area, canvas dams erected. These matters are put in hand at high speed; but the water cannot come through in a minute. As soon as possible those tough river boats with their heavy pumps were in position, hose had been flung across the mudflats, powerful hose-laying lorries were setting out their twin lines in the direction of the City danger zone a mile away. At the same time, mobile land pumps were seeking strategic positions by the riverside where there might be water within reach of their suction pipes. These pumps eventually operated at bridges and dock basins situated some distance from the fires. Before nine o'clock a message from the Guildhall reported that the spire of a neighbouring church was in imminent danger of collapse and might spread fire to the historic hall itself. Reinforcements were required here - and in a hundred other places too. By that time over three hundred fire engines had been sent to the City. More had been diverted to fight fires in other parts of London.

The fire situation in that square half-mile of the City called the 'danger zone' was assuming alarming proportions. Even in peacetime its narrow, congested streets flanked by warehouses filled with inflammable goods made the possibility of a conflagration in this area an ever-present anxiety for London's fire chiefs. Fires were started in hundreds of buildings and orange fireglow blazed with the bright force of sunlight. A glare rose high into the sky that could be seen from great distances beyond London. Dark city alleyways and passages, curtained for a century by tall wails, exchanged their twilight gloom for a flood of yellow light in one theatrical moment. Firemen walked the streets through blinding spark showers that drove down from the roofs with the intensity and regularity of a snow-blizzard. Waves of flame rolled across whole streets; black clouds of smoke smothered the air. Firemen fought on. It seemed that they fought a lost battle. And high-explosives were falling, killing and injuring men.

Three important City fire stations were burning and had to be evacuated. The controls retreated out of the immediate danger area and set up again on the outskirts of the fire. Then, towards ten o'clock the roof of the historic Guildhall caught alight. A control staff in the vaults stuck to their posts until the fire had all but reached their door, but eventually these, too, had to be evacuated. Before midnight the all-clear siren sounded: It was a relief to those working that the bombing was at an end; but civilians further overloaded lines of communication with what they thought to be new information as to unreported fires. By this time various units of the emergency water service were in operation and the supply of water was being strongly augmented as the minutes passed. Large canvas dams had been erected and firemen cheered as they saw the water pour in. Pumping operations gathered force. Firemen gathered new hope. Reinforcements of fire engines had arrived. So that now there was water and there were pumps and firemen to work them. Almost enough of each: enough anyhow to start effectively the stemming of that ferocious flood of fire. And more engines were racing on their way, more emergency water units were coming into operation every moment.

From then on every man was at work. Gutters ran with black water that streamed off the charred buildings. Dispatch riders scrambled their motor-cycles over the maze of snaking hose-lengths that littered every street. Petrol lorries arrived; and here it may be noted that some of these heavy vehicles loaded with inflammable petrol were driven by girls of the Women's Auxiliary Fire Service, driven through dangerous blizzards of spark and flying ember. Women also brought canteen vans into that inferno, working tirelessly on through the night to feed the thousands of firemen on the job. The work of feeding those crowds of men was a problem. What could be done was done. Yet many firemen there had to work the night through and on into the middle of the following day without refreshment or rest, A fifteen-hour stretch of this hard, wet work without so much as a cup of tea is no small order: but the men knew what they had to do and they stuck to it.

Perhaps the quality of the fire-scene that night may best be illustrated by the mention of a few individual incidents characteristic of what was going on throughout the City. One street was ablaze from beginning to end. Debris blocked both outlets. Trapped in this basin of fire were half a dozen or more engines together with the personnel manning them. For a long time these men sought for a means to defeat the spreading blaze. Both ends of the street were sealed. There seemed nothing for it but to abandon the engines. This decision was at last taken and the men looked for a way out. By the grace of God they found one - some stairs leading to the entrance of an underground tunnel. Via this tunnel they made a perilous escape, leaving behind them a grave of fire and collapsing walls.

Firemen tell the story of a fantastic incident in another part of the City. Once again fire stretched the whole length of a street. Fire engines were still positioned at the kerb, still able to pump up water on to the blaze. But it got too hot for one of the small trailer pumps. The pump caught alight, the men had to retire. But they found they were still gripping hose fed from the burning pump. Curiously enough the pump was still forcing out water - the fire had not yet upset the engine. So they turned the jet on to the pump - and there was realised the anomaly of a fire engine on fire providing water to save itself!

A great hospital was snatched from the flames after a long battle and a fortunate choice of tactics by the officer in charge. It was like this; a corner of the block of buildings that comprised the hospital faced converging streets along which the fire crept rapidly from house to house. It was apparent that, unless this advance was somehow checked, that particular corner of the hospital would eventually fall victim to the flames. Set in the basement of the hospital was a swimming bath. This contained 40,000 gallons of water. Not a big supply, but enough for an attempt to hold the fire off the corner of the building. With this end in view, one of the smallest pumps was manhandled down into the basement and its suctions lowered into the bath. This pump was to relay water to a dam out in the open. From there another pump could feed on the water sent up and direct its jets on the approaching fire. This it did. The men stood on the corner and played water on the advancing enemy as a machine-gun post might make a last stand against a ceaseless flood of attacking infantry. Those men mowed the enemy down. The fire was prevented from crossing the street. Forty thousand gallons and a laughably little pump saved the huge hospital.

Down in the dark waters of the dock basins there was trouble too. A fire boat had nosed its way into one dock at a time when water was urgently required. This boat was to lift an invaluable ration of water to feed relaying lines leading to the fire sector. But beneath the surface of the dock water lay disaster-a hidden wreck. As luck would have it, the course of the boat was directed across the wreck's position. There was a collision. And for a time, at a period when minutes were precious as life and inexorable as death, that fire boat was out of action.

Many incendiaries fell across the dome and amid the towers and buttresses of St. Paul's Cathedral. Frantically, the Cathedral staff clambered over the treacherous slopes of the high roofs in their efforts to deal with the shower of bombs. Three outbreaks of fire occurred. Creeping across dizzy stone escarpments, reaching out among buttresses that fell sheer to the street many feet below, they yet managed to beat the bombs and quench the burning timbers; but another danger threatened the great Cathedral. Many buildings surrounding her were ablaze from top to bottom. Flames reached out across the street, licking close to the venerable stone. Firemen on the ground and on ladders worked furiously. Hour after hour they slaved to hurl back that advancing wall of fire. But hurl it back they did and after a battle lasting hours they cheered to know that St. Paul's was saved.

The fire at the Guildhall was intense and terrible. But to the detached observer - the scene must have seemed picturesque despite his horror at such destruction. Flames trembled up and down the skeleton of the great tower that now flickered like a huge pyrotechnic display piece. The silhouette of giant roof-rafters stood out black and sombre against the red fire that was breaking them. On shallow flats that buttressed the main building the fly-size shapes of firemen could be seen desperately fighting a sea of fire. They were right up against eddying waves of flame a hundred feet high; it looked as though they were already in it. Certainly they worked in great danger of a general collapse. But they were not wholly defeated arid, though the building suffered harm, it is not irreparable and the flagstaff still stood to fly the Union Jack when they hoisted it in the fresh light of dawn. By 7.30 am all fires were under control. But throughout the next day and night the work went on. Fresh outbreaks had to be checked, smouldering debris and piles of hot white ash doused of their last vestige of fire. By midday many of the crews that had striven in the night were relieved by men returning from leave. And so to the drugged sleep of exhaustion after a battle that had lasted since seven o'clock on the previous evening. That morning a strange city greeted workers who sought their offices. Great blocks of building had been gutted to the ground. The charred shells of burnt-out shops and offices lined whole streets. Beautiful architecture had disappeared overnight or now faced the daylight black and broken - pathetic evidence of the new German culture. Historic churches designed centuries before by the master hand of Christopher Wren lay shattered, their very altars choked with the black mud of sodden ember and ash. And yet with all this melancholy toll of destruction the heart of the City, together with its great Cathedral, had been saved.

On that night of 29th December, fourteen firemen were killed and over 250 were injured. This in addition to casualties among the civilian and military help that was so readily volunteered. To these men and to the thousands of their comrades who fought with them went the gratitude of the people of London.

Your fireman will gloss over scenes of grandiose spectacle. He is used to them. Rather, he will tell you of a more intimate world turned topsy-turvy by heat and flames. He will tell you of the fire at that rubber warehouse when molten rubber covered the floors, when his boots stuck, firm at every step so that he felt like a fly on flypaper. Or of another sticky job at a toffee store, when a lava-like flow of freshly warmed toffee enmeshed his hose and brought again to his nostrils the old exciting smell, a hundred times magnified, of the toffee pan in his mother's kitchen.

Eccentricities will happen. A man who fought a fire at a paint factory found the next morning that his boots were beautifully varnished. In a certain church in a south-west district the best vantage point for the firemen was the pulpit itself: that night the men preached a sermon of water on to a congregation of flames.

Again, somewhere in Dockland a wharf housing whisky caught alight. A wide curtain of the precious spirit streamed down the outside walls from burst crates up on the top floor. As it descended the spirit caught fire. And thus the whole side of the building became one sheet of falling flame - as blue as the flame on a Christmas pudding. Once, too, a butcher's shop caught fire. By chance the flames did not actually consume the stock of meat. But they were near enough to give it a fine roasting. Firemen saw huge sides of beef browned to a turn. The air was sweet with the perfume of good crackling pork.

Your fireman will also remember less pretty stories than these. Tales of fires spreading towards ammunition dumps, fatal moments, when every ounce of energy must be exerted to stem the tide of flames flickering round drums of inflammable petrol. One motorman (driver of the pump) will tell you of the night he drew his water from a flooded bomb crater with a delayed action explosive full at the bottom. His heavy suction dragged the sides of the crater: he and his pump remained within three yards of the sleeping bomb for over two hours. He will tell you how much pure philosophy he learned during those long hours of waiting and watching.

Some of the men seem to remember the cold more than the heat. 'You can have the winter! Used to get drenched through the first five minutes. Drenched to the skin-under my coat, in my boots, everywhere. Then an 'outside' job. Standing outside the building with the cold wind cutting into my soaked spine. And that would go on for hours. Give me the middle of the fire any day...' Yet there is another man who looks with a more kindly eye upon the drenching question after a certain night back in January.' ...we heard the whistle and ducked for it. A flying brick or something must have hit our hose. It burst. The water gushed out everywhere. I was drenched through in a second. I cursed like I've never cursed before. Then I stopped suddenly for I saw something on the ground about eight inches from where my head had been. It was a huge bomb splinter - a square foot of it! Well, the water didn't seem so bad after seeing that...'

There are some interesting incidents in the journal of a fireman stationed near Streatham Common. 'A solid mass of roaring flame nine storeys high completely around the whole three sides of the dock...men carrying fifteen hundredweight pumps into the dock, over piles of hose, like slaves during the building of the Pyramids; 400 soldiers from the Tower of London to help roll the hose...' And. 'The water was hot, quite hot. It was being pumped into a building and as it gushed back into the bomb holes we pumped it back again.' And a third excerpt that speaks the mind of a pump operator down in the street below the fire '...Give them water, gallons, thousands of gallons of it. Don't fail; watch those gauges, easy on the pressure, a man's life depends on the way you get your water through, a man perched high on that building. A sudden jerk of your controls and he would hurtle to his death... the live thing in your control is only a machine, you're its master, make it behave...'

Tales of rescue and tales of tight corners abound. Many have been retold in the Press, many live silently in the memories of those men concerned. These and a hundred other things fade into the maze of flame and hot smoke, long hours and heavy equipment, noise and burning, much water and little sleep that goes to make up the fireman's particular blitzmare.'[22]

In the first six months of 1941 the London Fire Service dealt with five large-scale attacks. 'One night in May 1941' [10/11 May] continues Annette Coppard, 'the worst raid of all on London killed and injured more than three thousand people. Over two thousand fires were started. From ten miles away we could see London burning. As we looked at the red glow in the sky over the capital we wondered if there would be anything left of London at all. It took almost two weeks to put out the last of the fires.

James Negley Farson, an American author and adventurer who reported in Britain for the *Chicago Daily News* and Tom Purvis made a record of the raids in picture and print and what the Londoners were doing during these 'shocking incidents'.

'A year ago the bombing raid had been a vague terror' wrote Farson. 'Now it had been experienced and measured and, as H. G. Wells wrote: 'Our remarkable people seem to enjoy being at last, all of them, under fire. They feel that they are doing their bit and no longer being 'protected' by the men in uniform.' Even more, since Wells wrote that, we have seen letters from soldiers in camps in the West Country and Scotland, actually pleading to be allowed to come back to London. They did not want to feel safe, they insisted, when they knew their wives and children were in danger. But since Wells wrote that, on September 15th, 1940, there has been a change in the temper of London's people. It does not mean any diminution of courage. On the contrary, from what Tom and I saw and heard in the various shelters, down Thames River and among the debris of bombings, we believe that the courage with which the Londoner carries on - carries on, mind you - is much more to be admired than the original instinctive gutfulness with which nearly everybody in this amazing city reacted to the first downpour of disaster.

'For, at the beginning - say, the first ten days after September 7th - the average Englishman was far too busy to think very much about what he would do about anything. Having no alternative, the 'little man' and his wife just carried on. The A.F.S. people, the air-raid wardens, the police, the women ambulance drivers - all the great mass of auxiliary and regular city services - carried on because that was their job. They had a job to do - with a great proportion it being the very thing they had been twelve months in training for - so they did it. It was automatic. It was poignant, in some ways, in its individual, instinctive heroism - because, certainly, a fair number of these brave people, especially the fire-fighters, felt, at certain moments that they were going to be killed. (Lots of them were.) But it was not a premeditated act. I think it is true to say that in those first startling nights very few of the people involved in the actual fire-fighting or rescue work made any conscious call upon their courage. They just met it.

'And that marks the changed temper of London. Because now it is conscious courage. Wells could not write now that 'our remarkable people seem to enjoy being at last, all of them, under fire'. The early exaltations (because these bomb raids certainly did produce them!) have died. In their place has come a grim prospect. For a vast proportion of the London population it means not only the mathematical proposition about not getting killed, but the dead certainty of a wretched, uncomfortable night. Probably a whole winter of dreadful nights. Sheer physical pain. When the Londoners steeled themselves to carry on, resolutely facing such prospects, that is possibly the greatest and finest spirit which has been born among these ruins in London... The Englishman's calculated courage which has made him decide that he is going to stick it out, really does mean that London can be made unbeatable... The plain fact seems to be that the raids can never be made so heavy that they can break this determination. And the story of London now seems indicated to be the grim struggle of physical resistance. The die is cast. Having failed to panic the Londoners in their first furious air raids, the Nazis can never panic them.

'...Then I recollected that the previous night, when the wind shrilled around my house, I had no longer felt the pleasure of oncoming autumn weather. It was not the invigorating days and crisp night's drawing on. I could not curl myself up in my bed in a cosy defiance of what was without for the Banshees had gone off. A roum-roum-roum of Nazi bombers was overhead. Bru-mp! bru-mp! went the guns. There were sinister cruu-mps from over Chelsea way as the bombs exploded. For so many nights had I lain listening to Chelsea getting it (where the Germans were trying to get the bridges, but missing them), that I had begun to be surprised to find that all my friends were alive every morning. (I write this to-night with a Nazi bomber dropping his death only a few blocks away. A whistler has just come down as I write these words.)

'A stranger spoke to me:

'Last night we did get a packet!' he said. 'Have you ever tried to breathe scalding water? Scalding water...when you're in a ball of black cotton wool...with the flashes of the bombs going off like that around you?' He made several dabs with his hands, to show me the sudden explosions of bombs going off in all that murk. 'Well... that was what it was like.'

'This unshaved young man, in dirty brown dungarees, who had been smoking his pipe so quietly that he did not appear to be thinking about anything, was, he told me, a barge-tender.

'We had to move them,' he said, poking his pipe towards an island of barges anchored in mid-stream, two of which were only being held up by the others.

'Behind the barges two fire-tugs were still throwing heavy jets into a smoking warehouse.

'Lumme!' he said suddenly; 'those AFS men are marvellous!

'We had eight fires in the docks last night ... and they had 'em all under control by daybreak this morning.'

'In some parts of the river it is hard to tell the dilapidations of time from the wrecks caused by bomb explosions. But when you see the roofs curving over into the water you know a bomb has been there. A wharf on the other side of us had been hit and its insides were hanging out. They were wool... and it looked like the stuffing of a doll hanging out.

'In these early weeks of the heavy bombing raids there was a certain dream-like quality to the life on the land. You just didn't believe it. One reason was that unless you were in the vicinity, you didn't see it. And one girl who I saw step out of a bus to enter her just-bombed house in Bramerton Street, even stood there and exclaimed to me, 'It just doesn't seem possible - does it? Had she come off the previous bus, she would probably have been dead.

'A big black ship, with a Dutch name, lies close to a brick wharf with half a dozen cranes bending and dipping for her cargo; part of her dun superstructure has been carried away - 'It was a time-bomb,' said the barge-tender. 'Took seven tugs to bring her in; but they're making repairs on her now as she unloads.' Here and there a small boat is anchored off the wharves, bearing a green flag with the white letters WRECK.

'It's over a barge, or something, sunk,' says the tram conductor - 'to mark an obstruction, you see.'

'And the parent ship of these 'WRECK' small craft has the remains of a Spitfire or a Hurricane (which it has apparently fished out of the river) dangling from her forward boom. A nasty sight.

'Hope he bailed out!' says the barge-tender.

'The barge-tender, staring at this river that he now knew, somewhat vaguely, he wanted to protect, watched me suspiciously as I pointed out its oddities on this bright and crisp September morning. The water, for instance, was a sullen, inky black, such as you will see running away from a burned building, which was exactly what it had been doing. There was a considerable amount of charred timber-planks, baulks and scantlings - floating along the tide-rip at the edge of slack bends. Not to be outdone, he pointed to one beam, a good foot square and some twenty feet long and confided: 'See that, mate? ...well, a bomb blew two of them up on the top of that warehouse over there.' And, encouraged by my open mouth, he sang out to a man standing on the pier to which we were tying up: 'Hi, mate, how'd you find last night?'

'Oh, we all just die here!' rang out the cheery reply. 'Cannons to the right of us, cannons to the left - Gor blimey!'

'I have never thought of the English as a revengeful nation - a conquering race seldom is - yet one of the most menacing things for Hitler in this characteristically quiet boat conversation was the way that everyone present dispassionately discussed the urgent need for the immediate bombing of Berlin. There was no false sentiment; it was just that no one there believed there was any other answer to the indiscriminate German night-bombing than to bomb Berlin off the map.'

In the three months from September to November 1940 there were 36,000 explosives dropped on London and 12,696 people killed in the capital alone. Yet proportionately more people were killed in Hull than anywhere else, whilst there were more air raids on Great Yarmouth than on either Birmingham or Liverpool. People the length and breadth of Britain who tuned into their wireless sets each evening were no doubt gratified to hear on the news and specially recorded dramatic interviews with aircrew that Bomber Command was giving the German cities 'a taste of their own medicine'. One August night 'a Pilot Officer of a Heavy Bomber Squadron' talked of 'Gate-Crashing a German balloon barrage.' Another night a sergeant wireless operator air gunner talked about the 'Bomber' that 'Shot Down Three Enemy Fighters.'

'We had a bit of excitement a few nights back' said the Pilot Officer 'when we ran slap into the middle of a German balloon barrage. Our luck was in. Not only did we get away with it, but we brought one of the balloons down. Our target was a synthetic oil plant at Gelsenkirchen in the middle of the Ruhr. It was a dark night - very dark - and we had come down to about 6,000 feet to find the target. We dropped our flares and located and bombed the works, then we climbed and went back to see what results we'd had. My second pilot was flying the plane. I'd been down in the bomb aimer's position, doing the bombing. Suddenly I saw a long dark shape silhouetted against the clouds; then, as the searchlights played across them, I saw three more. They looked rather sinister and they were on the port beam and port quarter about a hundred yards away. By now I'd gone up from the bomb aimer's position and was standing beside the second pilot. I gave instructions to the gunners to open fire at the balloons and we started to turn away to starboard to get away from them. Immediately afterwards the second pilot threw the aircraft into a very steep right-hand turn for he'd seen another balloon coming straight up in front of him. It had loomed up out of the darkness dead ahead and our wing tip just caught the fabric. If the pilot hadn't yanked the aircraft over quickly we should have flown right into it, the envelope would have wrapped itself round the plane and that would have been the end of the trip, but all that happened was that the aircraft bucked a bit, then there was a terrific explosion which we could hear even above the roar of the engines and I imagine the Germans were minus one balloon, though we couldn't see what happened.

'After the explosion, when we climbed up higher, we found we'd been flying along a row of balloons right in the thick of 'em. It was pretty amazing that we hadn't hit a few more, for when we'd been bombing, we must have been among all the cables. I knew there were balloons in the area-we'd been warned about them before we started-but the only way to find the target was to come down fairly low, so we had to take the odd chance. When we examined the aircraft the next day we found it hadn't been damaged at all.

'Another raid which I shan't forget in a hurry happened just before this balloon incident. On this occasion we were bombing the railway marshalling yards at Hamm. There's an important railway traffic centre here and it seems to be selected as a target most nights in the week. When we took off, the weather was pretty poor and at 7,000 feet it was freezing. Over the North Sea we struck heavy banks of cloud. I climbed to 14,000 feet, but even at that height we couldn't get out of it - so we just carried on flying through cloud; there was nothing else we could do about it. When we were about fifteen minutes away from our target, the port engine began to splutter and the engine revolutions dropped. This time again I was down in the bomb aimer's position preparing the bomb sight. I realized that we'd probably got ice in the carburettor, so I came back to the second pilot. The starboard engine spluttered and soon both engines ceased to give any power at all. Our air speed indicator packed up, so did the altimeter. We didn't know whether we had flying speed or how high we were. The second pilot and I were flying the machine between us while he was fixing the warm air control which had become disengaged. This is an arrangement by which the air, instead of being sucked straight in, is warmed up by the heat of the engine before being passed through to the carburettor. He'd got both his hands on the warm air lever, forcing it down as far as it would possibly go and he'd got his feet on the rudders while I grabbed the stick, keeping the aircraft on an even keel. I could tell by the feel of it that we were going down very quickly. We were heavily iced up; the wings had a thick layer of ice on them and one couldn't see through the windscreen because that too was covered in ice. It didn't matter very much about that because it was so dark and we were still in cloud, so we wouldn't have been able to see anyway. I decided that if the engines didn't come on within another four or five seconds, I'd give the order to abandon the aircraft. I'd got the words on the tip of my tongue, when the port engine spluttered a couple of times and began to pick up again. It would still have been impossible to maintain height with the amount of ice we had on the aircraft, but I decided to hang on a little longer before giving the order to bale out. The starboard engine picked up and after a bit more spluttering, both engines started working normally again.

'We flew on for another half minute or so and then the altimeter started registering. I looked at the height and found it was approximately 4,000 feet, that meant that we'd come down in a dive about 11,000 feet. We were just recovering from this when we ran into an electrical storm. The effect was so weird that I began to wonder whether we hadn't arrived in another world. The others said afterwards that they began to think the same thing too. Everything seemed outlined in a blue haze. The propellers made shining spinning circles. The two guns in the front turret were pointing up in the air and there was the same blue haze round them too. The front gunner reported that there were sparks jumping from one gun to another. The rear gunner afterwards said, that for a minute he thought his guns were actually firing and he couldn't understand it. As I looked at the second pilot's face, I saw that it was ringed with blue. The tips of his fingers had the same blue haze around them. It covered the instrument panel and ran along the leading edges of the wings. That lasted about two minutes. It was one of the weirdest experiences I have ever had. I was very glad when we got out of the storm.

'We flew on and arrived over our target area. The cloud was still so heavy that it was impossible to locate the marshalling yards, so we turned and came back and on the way we bombed the alternative military target which had been allotted to us. Three fighters picked us up near Rotterdam. First of all the rear gunner reported one enemy aircraft apparently trailing us, then he was joined by a couple of his pals. We got all set for a bit of a scrap but nothing happened. They didn't attack and we arrived back at the base without further incident.'

The sergeant wireless operator air gunner, who was recently awarded the Distinguished Flying Medal for gallantry in operations against the enemy and who was from Derby, began his story: 'It was on the way back from a raid in the Ruhr that these three fighters had a go at us. We had been flying for about a quarter of an hour after bombing our target when we were picked up by searchlights. I called up the pilot on the intercommunication set and told him that the lights were dazzling me. They held us right across the town of Wesel, which is to the north of the Ruhr; then, on the other side of the town, the pilot finally got out of them. There was no anti-aircraft fire, so I was keeping a particularly sharp look-out for fighters. Suddenly, tracer bullets started flying past the turret and I saw three fighters coming in at us from the rear. One was coming in from the starboard quarter and below us; the second was above and practically dead astern; and the third was five or six degrees to port and he - like the one on the other side - was also attacking from below. To me it seemed that all three were converging on the rear turret. The one on the starboard quarter seemed to be pretty close, so I had first shot at him. The first burst seemed to hit. If you can get your first burst all right, you can usually guarantee to get your following ones in too, unless things are particularly awkward; so I just kept pumping quick bursts into him - six or seven altogether. He was hitting us too. Some of his shots went through the tail plane, the rudder and the wireless mast and an explosive shell from his cannon hit the armour plating of my turret. 1 didn't realise at the time that the shell had actually hit us. I thought it had exploded just outside. Anyway I know the bang deafened me for thirty-six hours afterwards.

'The fighter got to within about one hundred or a hundred and fifty yards of the rear turret; then he pulled up like an aircraft pulling out of a dive. He seemed to hang there for a bit and I got in a few more bursts right into the belly of the machine. I saw him turn over and then I swung the turret on to the second fighter which had been closing in all this time, firing his four guns. I could see four streams of tracer coming at us. Out of the corner of my eye I noticed the first fighter go down in flames. He exploded in the air or when he hit the deck - I couldn't say which.

'This second aircraft was the one which was flying slightly to port. I missed him with the first three bursts, because I was misjudging his speed, but the fourth burst hit him all right and after that I just kept repeating the performance. He was pretty deadly, too and did further damage to our plane. The navigator got hit in the leg - not badly though - but nobody else was hurt. Then the fighter curled away out of my field of fire and that was the last I saw of him, but the second pilot said he saw him go down out of control. After this the third enemy fighter came down on us. He closed in to about three hundred yards but wouldn't come any closer. I got a bit fed up with this so I fired a good long burst in his direction and he sheered off. We didn't see him again. Altogether, I've done just over twenty raids over Germany, but that was the most exciting one of the lot. I've got my twentieth birthday coming along in a few days' time and I hope to be over Germany that night.'

The fact that the Luftwaffe was unable to knock out RAF Fighter Command in the Battle of Britain owed much both to the workers of the Spitfire factory in Castle Bromwich, Birmingham and those at the BSA who made the Browning machine-guns for these fighters. The second largest city in the United Kingdom with around a million citizens in the first half of the twentieth century, Birmingham was renowned as the city of a 1,000 trades.

Because of its size and manufacturing importance, both its people and factories played a vital role in the British war effort and by 1944, 400,000 Brummies were involved in war work. This was a greater percentage of the population than anywhere else in the UK. By the end of the war, Castle Bromwich was producing 320 Spitfires and 20 Lancasters a month - more aircraft than any other factory in the UK. Elsewhere in the city at the Longbridge shadow factory, men and women turned out 2,866 Fairey Battles, Hurricanes, Stirlings and Lancasters; whilst at the nearby Austin works almost 500 army and other vehicles were made each week - as well as a multitude of other goods. Indeed, the array of war work in Birmingham was staggering. Bristol Hercules engines were made at Rover; Lancaster wings, shell cases and bombs were manufactured at Fisher and Ludlows; Spitfire wing spans and light alloy tubing were produced at Reynolds and plastic components came from the GEC. Up to the Battle of Britain, all the aero-carburettors for the RAF's Spitfires and Hurricanes were made at SU Carburettors - and if it had been destroyed the air force would have suffered a mortal blow; whilst Serck produced all the radiators and air coolers for these fighters. Wolseley and the BSA (Birmingham Small Arms) in Small Heath all strove hard for victory and when the BSA was hit badly in November 1940, Churchill himself was alarmed at the consequent national fall in the making of rifles.

The Luftwaffe's air raids on Birmingham had begun on 9 August 1940 when a lone German plane dropped bombs in Erdington. One of them killed Jimmy Fry, an eighteen-year old who had been on leave from the Army. Four days later the Luftwaffe hit the Castle Bromwich aircraft factory, killing five workers; and on 15 August bombers targeted the working-class areas of Small Heath, Hay Mills and Bordesley Green, where a bomb killed Mr. and Mrs. Wall, their daughter Dora, Mr. Wall's 79 year old father and three friends of the family. Then on the night of 19 August bombs were dropped on a thickly-populated working-class district. In one house 65-year old Elizabeth Smith was killed with her four year old granddaughter, Joyce. One hundred yards away two schoolboy brothers were killed. Their mother received a fractured arm and cuts and their sixteen year old daughter had severe lacerations. She died in hospital. Her father was at work, unaware of the devastation of his family. This night raid was followed by a bigger one on 25 August when 50 bombers attacked Birmingham. 25 people were killed. They included Robert King, who was in the Auxiliary Fire Service, his wife, Maude and their fourteen-month old baby, also called Robert. The next night, attacks were made on the Snow Hill neighbourhood, distinguished chiefly by its houses, railway station and warehouses.

On 27 September Fort Dunlop was attacked in daylight. Miss B. J. Wright was a junior clerk at the works and was at the bus stop with other employees waiting to go home at 5.30 pm when 'a German plane dived out of the clouds (no warnings) and opened fire with a machine gun. He was so low we could see him laughing in the cockpit. It was pouring with rain and we all dived for the gutters flat on our faces.' This shooting at civilians was not an isolated incident. On 14 October, Clementine Churchill, the wife of the Prime Minister, went to a Birmingham neighbourhood that had been damaged by bombing. She met with Mr and Mrs Hartle, whose home was in ruins. They told her that 'our house is down, but our spirits are still up'. That evening German attacks on Birmingham resumed. The next night, 59 people were killed in more severe attacks. The Germans dropped over one hundred high explosive bombs, several delayed action bombs and hundreds of incendiaries. On 24 October, shortly after midnight 189 major fires were blazing. There were deaths and damage across the city centre and a shelter in Cox Street, Hockley, was hit and 25 people were killed. The next evening the Carlton Picture House in Sparkbrook was bombed. The manager had told the customers to sit in the stalls beneath the balcony of the circle. This gave some protection but nineteen people sitting in the rows nearest to the screen were killed. [23]

Though the German defences became so much stronger during the winter of 1940/1941 they were by no means negligible during the summer and early autumn of 1940. And they were being strengthened all the time, especially in the places most frequently bombed; at the beginning of August; for example, a great intensification of anti-aircraft fire in Hamburg was reported. As early as July 1940 Winston Churchill found time to address the Chief of the Air Staff in these words: 'In case there is an attack on the centre of Government in London, it seems very important to return the compliment the next day upon Berlin. I understand you will have by the end of the month a respectable party of Stirlings ready. Perhaps the nights are not yet long enough. Pray let me know.'

Following raids by German bombers on London and other English cities on the night of 24/25 August, the War Cabinet sanctioned the first raid on Berlin the next night, 25/26 August, when 103 aircraft were dispatched on operations and approximately half of these, mostly Hampdens and Wellingtons, were sent to Berlin. A formidable concentration of guns and searchlights was ready for the bombers. The Hampdens were at the limit of their fuel capacity in the thick cloudy conditions and three were lost. Three more were forced to ditch in the sea on their return flight. When RAF bombers were next over Berlin on the night of 30/31 August the flak was even more intensive than during the first attack and there were great concentrations of searchlights.

By the autumn RAF Bomber Command was raiding Berlin, Bremen, Hamburg, Wilhelmshaven and Essen. A Sergeant pilot described the raid on Stettin on the night of 14/15 October on the BBC.

'Last Tuesday we were called in to be briefed and told that our objective was the synthetic oil works at Pölitz, near Stettin. That meant something more than six hundred miles out and six hundred miles home. Roughly 1,300 miles for the round trip. When they briefed us - that is, when they told us all about the target, how to get there and what to do, it was pointed out to us that this oil plant was able to produce a million metric tons of fuel for the enemy every year, so long as it lasted. The intention of the raid was to make the oil last a very short time and even though it means cutting the story I think I can say that that intention was carried out. The raid lasted about two hours and I don't think the oil plant lasted as long as that. It was a blazing mass when we left.

Our Intelligence Officers told us that we should be able to identify it because of its position near the river and because it had six very tall chimney stacks clustered together at the south-eastern end. We were to attack in the neighbourhood of those chimneys.

'The night was so clear and the country so plain below us that we could map-read our course and pick up first one landmark and then another. I have been over there a few times and so know much of the country, so does my navigator. We had the moon and the luck with us and could see everything. There was a fair amount of anti-aircraft fire at Cuxhaven, but there always is. We came through that all right and flew on our course, striking the river that was our guide to the target and avoided Stettin because we were on a particular spot and the town meant nothing to us, then. Following the river in good moonlight, we realised that some other members of my squadron had been over the target before us. There was a lovely fire blazing which we could see for the last fifty miles. It was blazing away like the Fifth of November. It looked as though there would be nothing left for us to do but there was. We came in from the north and the wind was blowing fairly freshly from the south. Until we reached the neighbourhood of the target we were flying fairly high. Then we began to glide and losing height went in to have a look. At 7,000 feet we ran into a pall of smoke. Somebody had already hit the target a very good crack with their stuff. The place was well ablaze and smoke was coming up in thick, billowing waves. But through it all we could see just what we had been told to look for: the chimneys of the power-house, with this difference. We had been told that there were six chimneys. When we arrived there were only four. Somebody else had brought the other two down, but those four stood up, one of them slightly bent, like the crooked fingers of a maimed hand. I decided not to attack from there and circled round out of the smoke. Below me lay the target, blazing merrily. This was a real military objective, not just a row of houses. There was a works building that I should judge to have been nearly four hundred feet long. It was two storey's high and the flames were pouring from the windows. They went out into the open like great flashing tongues and came in violent gusts. Something inside that building was feeding the fire every second and red flames and black smoke belched out without pause. Then I saw the four chimneys that still stood and decided that that was my individual bit of the target. We came in and bombed in two sticks. The first stick was high explosives followed by incendiaries and it went straight across the target. There was no doubt of that at all; although I, being busy with handling the aircraft could not see the results then, the tail-gunner saw what happened and reported that new fires had broken out, after good explosions.

'Then I turned round and came in lower, to drop my heavier bombs. When the observer said 'bombs gone' I circled at once and saw two of the chimneys buckle up. It was an amazing sight and very hard to describe, but those chimneys went down straight for a while and then fell over sideways, as though they were sinking to their knees. Then they toppled over on their faces. They were big chimneys and they fell into the heart of the fire, which spread rapidly like a red sheet on the ground. The tail-gunner reported that another fire had broken out and then an anti-aircraft gun started on us, just one and it was wide. But there was plenty of anti-aircraft fire over Stettin, where we were not. Searchlights came up too and the tail gunner opened up with his machine-guns, from 4,000 feet. Two searchlights went out. When those chimneys had melted into the fire and our bombs were all used, we dropped our flares as incendiaries, too. We didn't need to bring anything back and flares can do some damage if they fall well. We left the target like an inferno. Flames were obscuring the ground; we could see the fire but nothing else. Smoke was filling the higher stretches of the sky, with a bright red glow lighting the underside of the cloud. If they put that fire out they must have performed miracles, because I have seen some good fires on various trips, but none to touch the blaze we left at Pölitz. The tail gunner was reporting every little while and he was still reporting the fire when we were a hundred miles on our way home.

'We had a good run back, came to base on time and met the others who had been on the same target. I think we might say production was definitely stopped and won't start again for a very long time. It was a good trip.'

'If a bomber crew are to be successful in all they undertake' recalled an Irish sergeant pilot who was recently awarded the DFM, speaking on the BBC, 'it is essential that they should work as one man. My crew are an excellent team and that is one of the main reasons why we were able to pull this attack [on Berlin] off satisfactorily. When I was at school I was often told that if an Irishman, a Scotsman and an Englishman lived together in one room it would not be long before they fell out. I am glad to say that this does not apply to my team, perhaps because there are two Irishmen to keep the peace.

'So much depends on the navigator that it is just as well that I should tell you straightaway that it is he who comes from the same country as myself. The rear gunner is the Scotsman and the wireless operator the Englishman. The rear gunner and myself are more or less RAF 'veterans'. We have both been in the service about five years. The navigator joined up straight from school and the wireless operator gave up his job as a clerk to undertake what he calls 'more exciting work'.

'This was my first official visit to the German capital. I was over it once before, but that was after I had been attacking a target at Stettin. Afterwards we all thought it would be rather fun to make the Berliners go underground, so on our way home we flew over the city and made the ground defences waste a lot of energy and ammunition for nothing. But the flight was a great deal more thrilling than the raid on Stettin. This time on our arrival over Berlin we ran into a fierce barrage; shells were bursting all over the place, but in spite of this we spent about forty-five minutes over the capital before we dropped our bombs. We explored the city thoroughly and eventually found the target we were after. All the time I was manipulating the stick, the navigator was busy getting a decent pinpoint, while the other two members of the crew were giving me advice on which way to go in to avoid the ack-ack. We were then fairly high up, but the shells were still bursting pretty close to us. None of them actually rocked the aircraft, but two were close enough for us to hear them burst. There was a slight ground haze over the city but the moon penetrated it and showed up all we wanted to see. Suddenly, through the intercommunication system and above the roar of the engines, I heard the navigator say: 'I am sure that's the target.' Having complete confidence in him, I had no hesitation in shoving the stick forward and the nose of the aircraft down. Just before we went down I said to the crew: 'All right, down we go,' and just as we started I thought of my crew hanging on for dear life. We had been talking about dive attack for at least an hour before we got to Berlin. During the dive, which was made at a good speed, I had the target in the gun-sight and I held it there. All the time the target was getting bigger and bigger. Then I shouted 'let them go' and the bomb-aimer pressed the button. As soon as the bombs burst all the anti-aircraft guns opened up and every ten or twelve seconds we felt the most colossal bumps and the machine was jockeyed about all over the place. At first I started to climb, then to avoid the shells I had to dive again, then climb, then go from side to side, then do stall turns, then up, then another dive. This business of going up and down went on five or six times. One thing I am certain of is that I wouldn't dare to throw the aircraft about in daylight as I did that night.

'At one moment I saw a balloon go up in flames. Fire from the guns on the ground must have hit it. Actually I did not see the balloon until it caught fire; there was a flash and the whole thing was ablaze. We were only thirty yards away at the time and the cable down, whilst some of the burning fabric was sliding right in front of us. One of the chaps said that it reminded him of the Indian rope trick. In the end the cable fell clear of us and we all thought afterwards what a good thing it was that a shell had hit it. Twenty minutes later we were right out of the barrage and setting course for home.'

Included in the Bomber Command Battle Order was 311 (Czech) Squadron at East Wretham in Norfolk which was equipped with Wellington bombers. Flight Lieutenant J. Snajdr, one of the pilots on the squadron told the BBC:

'Up to this present time I am already on operations eight times against the enemy and now I am looking forward to many more missions. I am altogether nine years a pilot. When the Germans enter Czechoslovakia I escape to Poland. I am fifteen days only in Poland and then I come to England and here I am told that I should proceed to France with a group of twenty-five Czech airmen. I am wanting there to join the French Air Force, but there are difficulties. So instead I make application to join the Foreign Legion. All this group of twenty-five join also in accordance with advice from the Czech authorities in Paris and we go to Africa. At first we are employed as workers building highways. This was very difficult in the climate which is quite unusual for us. Then, after one time, we start with normal French drill infantry so that when these six months are over we can be regarded as French Legionnaires. During this time, however, the Czech authorities in Paris have made the necessary steps for us to join the French Air Force and when at last they have succeeded in this all these Czech airmen in the Foreign Legion are sent to different stations in Morocco. I spend another two months as pilot in the French Air Force in Africa; then I am sent to France and then when France collapses I am coming to England. That is in August. When I am here it is not necessary for me to make personal application to join the RAF, because everything has been arranged for us beforehand and so, after a little time, I join the Czech bomber squadron. In these eight missions, which I have made with them I am sent to eight different places, Berlin, Bremen, Hamburg, Wilhelmshaven, Essen, all in Germany and to these French Channel ports of Dunkirk, Calais and Le Havre. Three times I go as second pilot and then I am promoted captain of aircraft.

'The raid at Hamburg is my most successful, I think. This raid was made on a full moon night [on 21/22 October]. We succeeded in making an approach almost unobserved with complete silence on the part of the German guns on the ground, but so soon as we drop our flare, then immediately this anti-aircraft defence starts most violently and there are many searchlight also. I will never forget these thirty minutes which we spend over Hamburg simply trying to find the target in the docks. All the time they are firing at us very much. Then mine observer tells me 'OK. Now I am bombing.' He tells me it is very good, his bombing. I see a big fire start and then I hurry for England.[24]

'At my first bombing missions I have felt some excitement, but now I am quite accustomed to it and I do not feel any excitement at all. Upon one occasion [on 16/17 October] when I am second pilot we have to jump from our aircraft by parachute. To start with it was a very good flight this. We get to Bremen and we drop our bombs. Again these bombs are dropped at full moon and I am quite sure they are dropped on the docks. There is much anti-aircraft there also, but again we are not hit. In all these eight missions my aircraft is not hit. This time when we start to cross the Channel we are proceeding in cloud towards England and ice is starting to form on our aircraft. Because of these icing conditions our wireless set simply stops to work, so when we reach the shores of England we are without any guide at all. It is cloudy, windy weather, with complete darkness and we are about half an hour after midnight. It is so bad that at first we cannot recognise whether we are above the sea or above the land, though we have come down to less than five hundred feet. In the end we see the waves and the white foam and this enables us to recognise the shores. We are hoping that the wireless operator will succeed in repairing this deficiency in his apparatus, but it is not so. We remain in the air as long as fuel enables us to and then when we are still finding nowhere to land the captain is obliged to give this order to abandon aircraft. I am to go first. I shake hands with all of them and I go through the front hatch. We are now at two thousand feet. I make two somersaults and then I pull the rip-cord. I am wondering very much if this parachute is going to open. I have never jumped before. Then I have a feeling of the parachute coming out of its cover which is on my back and next there is a jerk when she opens and I start to swing in the air like a pendulum.

'It is raining very heavily and I am becoming soaked. As I descend, I notice a road. I shout, but apparently nobody is present. During this final period of descent I am prepared to land with my hands or my feet first, but unfortunately I first hit the ground with my face. I receive such a shock that it compels me to lie for some minutes to recover. Then I find I am in a meadow. I shout several times, but with no results so I fold my parachute over my arm and walk until I reach a house. When I knock on the door and shout again,

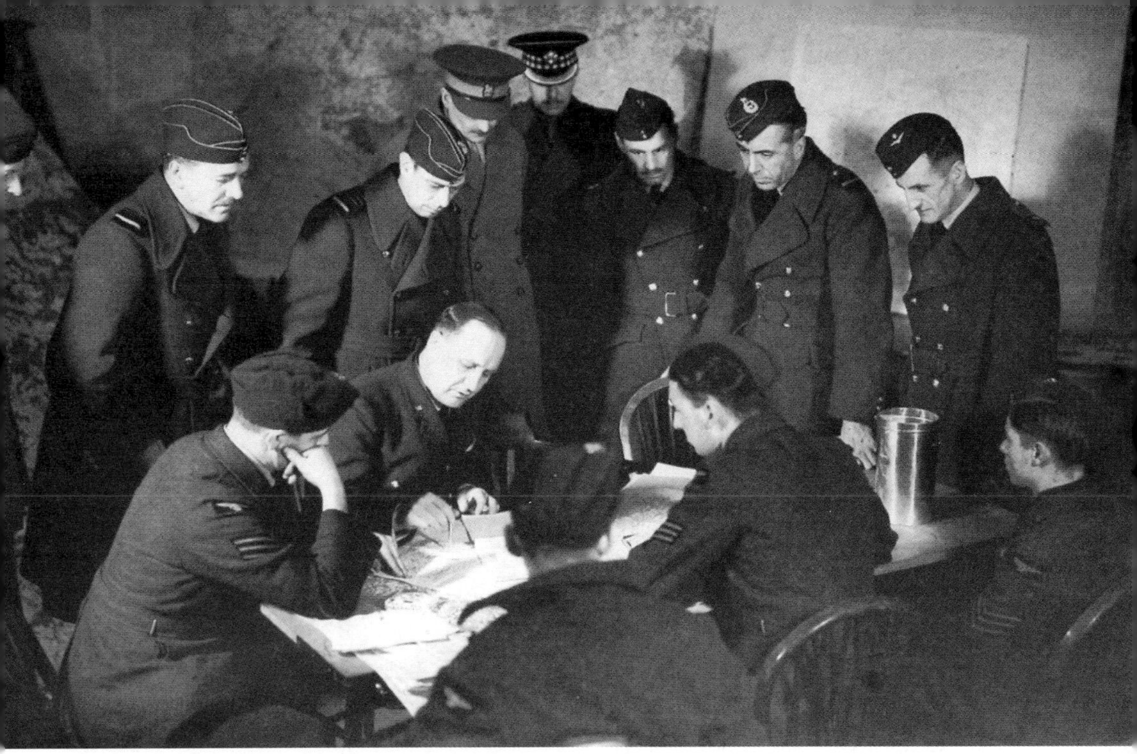

On the evening of 27 November 1940 HM King George VI visited 38 and 115 Squadrons at Marham, Norfolk, and stayed throughout the night to observe an actual operation in progress. In this photograph, the King (third from left), AVM Jackie Baldwin, AOC 3 Group (second from left) and Air Marshal Sir Richard Peirse, C-in-C Bomber Command (second from right), watch as members of a Wellington crew just back from Cologne struggle to concentrate during interrogation.

Air Marshal Sir Arthur T. Harris, having been recalled from the USA where he was head of the RAF Delegation, became commander-in-Chief of RAF Bomber Command on 22 February 1942. Harris was directed by Marshal of the RAF, Sir Charles Portal, Chief of the Air Staff to break the German spirit by the use of night area rather than precision bombing and the targets would be civilian, not just military. The famous 'area bombing' directive, which had gained support from the Air Ministry and Prime Minister Winston Churchill, had been sent to Bomber Command on 14 February, eight days before Harris assumed command. *(IWM)*

Above: Whitley Vs on 102 (Ceylon) Squadron at Driffield in Yorkshire, March 1940. The Whitley's slow operational cruising speed of around 140mph was not considered a handicap since German aerial opposition at night was virtually non-existent. These two Mk Vs of 102 Squadron were soon lost in action. N1421 DY-C was shot down by flak during an attack by six Whitleys on Oslo-Fornebu airfield, Norway on 29/30 April 1940. Flying officer K. H. P. Murphy and three of his crew were taken prisoner. LAC John Ellwood the rear gunner was killed. N1382 DY-A was shot down on an Augsburg raid, 16/17 August 1940 and it crashed at Walser Valley in Austria. All of Pilot Officer Mark Hubbard Rogers' crew were killed. The six Whitley squadrons of 4 Group were 10 and 78 at Dishforth, 51 and 58 at Linton-on-Ouse and 77 and 102 at Driffield. No other bomber squadrons operated the type, and when 4 Group expanded it was with Wellingtons and finally Halifaxes. (IWM)

Right: Thumbs up for a Whitley rear gunner. (IWM)

Below: Whitleys at dispersal. (IWM)

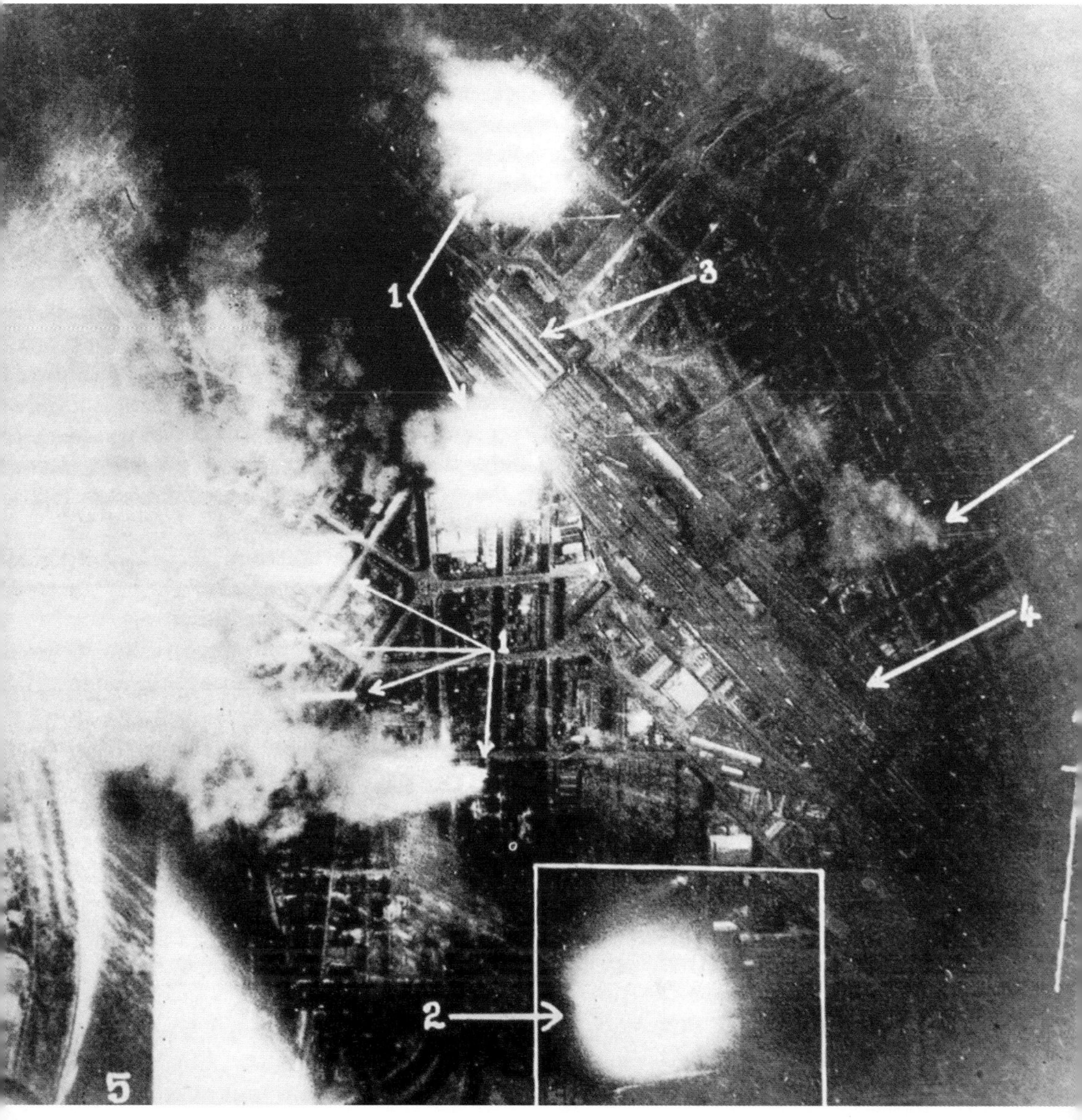

Operation 'Abigail Rachel' - Mannheim under attack, 16/17 December 1940. In retaliation for recent German raids, in particular the attack on Coventry, the War Cabinet authorised Bomber Command to target, for the first time, the centre of a city instead of specific industrial locations. The 134 aircraft sent represented the largest force yet dispatched to a single target. The central railway station (3) and marshalling yards (4) are visible in this view, with the Rhine on the left (5) and large fires burning (1 & 2). Nine bombers were shot down over Germany or crashed in Britain. The bombing itself was scattered and largely ineffective, but the raid was a portent of things to come.

The Avro Manchester endured an unhappy 16-month operational career with Bomber Command, thanks mainly to it being rushed into production with untried Rolls-Royce Vulture engines, which proved to be dangerously unreliable. 207 Squadron operated the type for most of that time, flying a total of 360 Manchester sorties with the loss of 25 aircraft.

The aircraft was the forerunner of the successful four-engined Avro Lancaster, which would become one of the most capable British strategic bombers of the war.

The barrage balloons protected many areas from German bombing; they were trundled out of their hangars by their solicitous crews, like so many nannies with prams, to float above the cities - chubby silver fish in the sunlight.

(Top) The Handley Page Hampden equipped the bomber squadrons of 5 Group for the first two years of hostilities. As they were replaced by types more suited to the Command's purpose, many Hampdens went to Coastal Command where the aircraft's long bomb bay was able to accommodate a torpedo. ATI37 UB-T of 455 Squadron RAAF is depicted here in May 1942 when based at Leuchars. The aircraft was damaged on 8 June that year and declared to be beyond economical repair. (RAF Museum)

(Middle) Three of the four man crew of a Hampden on 83 Squadron at Scampton in June 1940 exiting their aircraft after an operation.

(Bottom) Hampden I P1333 EA-F on 49 Squadron being bombed up with 250lb and 500lb General Purpose bombs at RAF Scampton in July 1940. P1333 and Sergeant M. G. P. Stratton's crew were lost on the raid on Merseburg on the night of 16/17 August 1940. All four crewmembers bailed out before the aircraft crashed at Alphen in Holland and were taken into captivity.

On 1 November 1940 207 Squadron, previously a training unit, re-formed at Waddington in 5 Group for the purpose of bringing the Avro Manchester into service. On 24/25 February the squadron carried out its first bombing raid when six aircraft were dispatched to bomb a *Hipper* class cruiser reported in Brest. One aircraft had undercarriage failure and crashed on landing back at Waddington.

For long-distance Nickel operations the Whitley squadrons used a forward base in France. On 30 November 1939 Czechoslovakian Flying Officer George Edgar Saddington MiD MC and crew on 77 Squadron flew N1357/KN-H to Villeneuve to refuel and then carry out a sortie, returning directly to home base at Driffield. (Flying Officer Saddington and his crew were lost without trace on a reconnaissance flight on 11/12 April 1940). These sorties proved bitterly cold for crew members, particularly the rear gunner, who often wore a fur-lined apron. This Whitley was lost after a Nickel flight over the Ruhr on 28 March 1940 when Flying Officer Trevor James Geach and crew crossed over neutral Holland and the bomber was shot down by a Dutch Air Force Fokker G1A. Sergeant James Emerson Miller who was Canadian, was killed and the rest of the crew were interned. *(IWM)*

Exhausted London firemen.

Danger UXBs was a hazard that had
to be endured long after the raid had
passed.

A London fireman
high above the city
dousing down the
flames after a heavy
raid.

Churches may seem the most dubious of military importance, yet they became primary objectives. Here the sculpted figure of Christ on the Cross hangs like some macabre burnt offering at the Church of Our Lady of Victories in South Kensington, London following a devastating German air raid on 13 September 1940.

'London can take it'... but so could other cities of Britain. Coventry's ordeal gave the Germans a new word in their vocabulary of horror, while the stonemasons of one of England's noblest cathedrals turned sadly in their graves.

Despite the devastation of the city of Coventry, life had to go on. There was work to be done, Germans or no!

Veteran American broadcaster Edward R Murrow wrote the following in an article published in early 1941; it could well have been about this picture selection.

These are pictures of a nation at war. They are honest pictures - routine scenes to those of us who have reported Britain's ordeal by fire and high explosive. These Englishmen have bought survival with their tender-roofed old buildings; with their bodies and their nerves. They offer you a glimpse of their battle. Somehow they are able to fight down their fears each night; to go to work each morning.

The pictures were selected with great discrimination. I would have shown you the open graves of Coventry - broken bodies covered with brown dust, looking like rag dolls cast away by some petulant child, being lifted in tender hands from the basements of homes. This spares you the more gruesome sights of living and dying in Britain today.

Of course, there are many parts of Britain untouched by the winds of war - little villages tucked away at the end of long fingers of salt water - greystone cottages at rest in the folds of the Cotswolds. Even in London there are miles of narrow, crooked little streets still undamaged. Much history has been blasted apart in Britain; but only the symbols are gone. The spirit of men who made Britain great still walks the streets in the shape of little men who think of themselves only as Englishmen.

Nickels being loaded aboard a 102 Squadron Whitley.

The Whitley was perhaps best known for its night leaflet-dropping flights over Germany, which began on the very first night of the war. A total of 123 'Nickeling' sorties were flown before Christmas, showering Germany and the occupied territories with propaganda leaflets, at a cost of only four aircraft. In the New Year, operations were extended to include such distant targets as Prague and Vienna, which were reached from forward airfields in France. The long leaflet flights over Germany and further afield exposed crews to bitter cold, with frostbite a constant hazard. Only the Whitley had the range to reach the furthest targets deep in occupied Europe, but such long trips proved to be feats of endurance for men and machines alike.

German flak guns

In and around Berlin, the Germans constructed massive towers for flak batteries.

there is a lady whose head appears from a window upstairs and I ask for help. This lady immediately vanishes and at once a gentleman appears in the same window with a gun which he points at me. When I see this gun, I am so weak already through all these events that I faint and next I find myself already in the house in an easy chair. When they have made sure of me they are very, very kind and they give me hot tea with whisky in it. The police and a doctor arrive with his car and he takes me to his house. He offers me a bath and pyjamas and a bedroom and something to eat and then I go to sleep. I am sleeping only for thirty minutes when the front-gunner arrives also. He explains that he had to hang for one hour from a tree with his parachute before succeeding to release himself and dropping to the ground. And soon after this incident we are back once more on operations. So now good night.'

A Flight Lieutenant on one of the squadrons who went to Munich, told listeners to the BBC that it was their first trip to the city. 'Our target was the railway locomotive and marshalling yards, almost in the centre of the city and only a short distance away from the famous Brown House of the Nazi Party. Just before we took off the senior intelligence officer came rushing over and said he thought that we might be interested to know that Hitler and some of his gangsters were to be in Munich that night to celebrate the anniversary of the Beer Hall Putsch of [8 November] 1923.

'Everybody was flat out to get there. They had included in my bomb load one of the heaviest calibre bombs that we have so far carried. I talked things over with the observer and we decided before we left that as the station commander had been kind enough to entrust us with the delivery of this heavy calibre bomb we'd go in as low as possible to make sure of getting the target. It was a beautiful starlight night and there was almost a half moon. We were checking up our course by the stars as we went out. Round Munich itself there was not a cloud in the sky. We passed an enemy aerodrome - all lit up for night flying - but on the way out we weren't wasting any bombs on that. We saw one of our fellows flying about five miles in front of us, getting a packet of stuff thrown up at him over Mannheim. He flew straight through it, but we turned away to the left and avoided the town. After Mannheim, Munich wasn't very far away and everybody was sitting up and taking notice. We were about twenty minutes' flying-time away when we first saw the flak and the searchlights coming up around the city. The navigator got a bit worried because we were ten minutes in front of our estimated time of arrival and he thought for a minute that we might have got off our course. Then we picked up a landmark - a goodish-sized lake-to the south of Munich - and set course from there. Some of the other fellows had gone on ahead to light up the target and we could see their incendiaries bursting.

'Flares were dropping all around as we went in. The guns on the ground were shooting quite well. I saw three flares shot down almost as soon as they had been dropped. We flew over to have a preliminary look at things and found we were about a mile south of the marshalling yards. We were low enough and it was so light that we could see houses and streets quite clearly. It was the bomb aimer's dream of the perfect night. Altogether we stooged round for about twenty minutes, checking up on our target. We saw somebody else drop his stick of bombs slap on the target. The explosions lit up the locomotive sheds. We came down lower and they were shooting at us hard. In the light of one of our own flares I saw a stationary engine in the yard. I could make out the glow from its fires and I noticed, incidentally, that it had steam up. We had to turn round and come back over the yards, making our run from south-east to north-west. Then we went whistling down. Tracer seemed to be coming up right under the wings and the bomb aimer said that he could see it coming up towards him as he lay in the nose of the aircraft looking down through his tunnel.

'All the way down in the dive I could see these big black locomotive sheds in front of me. The front gunner was shooting out searchlights, which I thought was a pretty good effort and the rear gunner was having a try at the same game, but it was more difficult for him. I'd told them that they could let loose with their guns and they didn't want telling twice. The bomb aimer got the target right in his sight. He said: 'I can see it: I can see it absolutely perfectly.'

'Then he called out: 'Bombs gone. I've got it.' As a matter of fact I don't see how he could have missed at that height. Both he and the rear-gunner saw the bombs burst. The rear gunner said that the heavy one made a dickens of an explosion. In the excitement I'd more or less forgotten that we had got this big bomb on board and the force of the explosion gave the aircraft a tremendous wallop. If we had come down any lower we should have been blown up. As it was we all thought we'd been hit. The effect was just as if a heavy shell had burst right under the rear turret. There was a stunned silence for a few seconds; then another babble of conversation when everybody decided that we were all right.

'We were still low down. Searchlights kept popping up. The front gunner put out two and the rear gunner put out four. It was a remarkable sight to see the coloured tracer going down the beams of the light. After that, it was a race back because we'd been told that the weather would close down over our base and that after two o'clock we'd be very lucky if we got in there, so we beetled back pretty rapidly. Altogether it was a perfect trip.'

On the BBC one evening in November a Flying Officer (he was on Hampdens) told listeners about the trip from England to Danzig and back on 10/11 November. 'Danzig, of course, is a port at the Baltic end of the former Polish Corridor and the journey was roughly equivalent to a non-stop flight from London to Madrid and back. The total distance, including necessary deviations, was something like 2,000 miles. Most of the credit for the flight belongs to the sergeant-pilot who was at the controls of the aircraft. For over eleven hours, without relief, he sat in his small cockpit, constantly on the alert and liable as captain of the aircraft, to be called upon for instant decision in any emergency which might arise. For us other members of the crew the strain was much less severe. When things were running smoothly we could take our minds off the 10b in hand and some of us, if we became too cramped, could get up and stretch our legs. But not the pilot. He had to stick it and, looking back, I still think that his staying power, matching that of the great engines which carried us on the long journey, was the most remarkable feature of our flight.

'When we were first told that Poland was to be our destination we were just about as 'pleased as we could be. Apart from the fact that the long flight promised some excitement, we were particularly bucked at the idea of cheering up the Poles by taking a crack at the invaders on their very doorstep. We'd already proved the idleness of Goering's boasts by repeated hammerings at Berlin and now this trip would give us the chance to show that we could not only elude his anti-aircraft defences in Germany, but fly right over them to help our friends on the other side.

'We set off in daylight. There were four of us in the aircraft: the pilot, two air gunners, one of whom was also the wireless operator and myself. I was the navigator and it was my job to guide the aircraft to the target and bring it safely home again.

'The sky was clear and the sun was shining when we started and it was still daylight when we crossed over the North Sea and into enemy territory. We were all keyed-up and keeping a sharp look-out for enemy fighters, but none came our way. Perhaps it was just as well for them that they didn't. We were determined that we were going to get to Poland that night and if anything had got in our way it would have had a warm reception. Our gunners' trigger-fingers were itching to go into action and our pilot, still fresh and alert, was ready to take any evasive action that might have been necessary.

'Actually, that part of our trip over Germany was so quiet and uneventful that it might well have been peacetime. But soon after the sun had set the weather changed. It became steadily worse and for practically all the rest of the outward journey remained thoroughly bad. As we flew further east we ran into a snowstorm and within half an hour the interior of my cockpit was some two inches deep in snow. It was fine and powdery and when I wanted to use my maps I had to blow the snow away before I could read them. The snow also managed, somehow or other, to get into my fur-lined coat, but as I was so cold by then, a little snow, more or less, didn't make much difference to me.

'Apart from the snow, we also had to contend with a particularly violent electrical storm and although none of the static penetrated into the aircraft I could see it striking the propeller tips and making the airs crews look like a couple of Catherine-wheels. I've never seen anything like it before - and I don't particularly want to again.

'Later on, as a change from the elements, we had a taste of enemy opposition. It came in the form of a certain amount of anti-aircraft gunfire, but it didn't really worry us. What it did do was to tell us that the enemy knew we were in the vicinity and it gave us rather a kick to think of the hundreds of air-raid sirens that were being sounded at every town and village along our course. We knew we were Poland-bound, but the Hun could have had no idea where we were going.

'We reached Danzig at last and just to look down on the Baltic Sea was enough to give us all a tremendous thrill. It looked so peaceful with the moonlight shining on the water and a background of cloud on the distant horizon. To do justice to the scene requires a poet rather than a navigator, so I won't attempt to describe it. Then I got on with the job of searching for our target. One thing I had to bear well in mind. If there was any possibility of my bombs falling on the civilian population then I must not drop them. As we circled the city unmolested by searchlights or anti-aircraft fire I spotted our target, a railway yard. Railway yards, whether they be at Hamm or Danzig, look much the same and as we came low the moon glinted on railway buildings and tracks and there, spread out before us, was the old familiar network. I told my pilot I was ready to bomb, but he wasn't taking any chances. 'You're quite sure of it?' he called out and only when I had reassured him did he straighten up the aircraft for the bombing-run. As we approached the target I took careful aim through my bomb-sights. The light was so good and the target so big that I just couldn't miss. The next thing I heard was a call from the rear-gunner. He had seen the bombs burst and they had started quite a large fire. It was obvious from the lack of opposition from the ground defences that we had taken the enemy by surprise. Or perhaps they had forgotten something we had remembered - that it was the anniversary of Poland's Day of Independence.

'After we had dropped our bombs on our enemies in Poland, we had the more pleasant task of delivering leaflets to our friends the Poles. Though none of us could read Polish, we had all studied the circulars with great interest before we set off. We were able to make out two passages. One of them was the news that President Roosevelt had been re-elected in the name of Democracy and the other was 'Long Live Poland!'

'And just in case there are any Poles listening to this broadcast, I'd like to end with the only Polish word I know-'Dobshal'- or, in English, 'It's all right!'

In the early afternoon of 14 November 1940 Blenheim crews in 2 Group in Norfolk were briefed for a night raid on enemy airfields. This night the Luftwaffe devastated the city centre, of Coventry. Plans of the large-scale attack were known in advance because of Ultra intelligence but the knowledge had to be kept secret from the Germans so no additional measures were taken to repel the raid. However, squadrons in 2 Group were directed to attack airfields such as Amiens, Etaples, Knocke and Rennes, from which the enemy bombers were operating to cause as much disruption as possible. Sergeant Jim 'Dinty' Moore a Blenheim gunner on Sergeant Roger Speedy's crew on 18 Squadron at West Raynham, recalls:

'At our briefing we were instructed to attack the aerodromes at Flers and Lesquin spending some time over their airspace to deter them from using their landing lights. At 1910 hours it was our turn to take off. We were soon climbing away from the airstrip and turning onto course. On this occasion the weather was exactly as predicted, so despite the attention of the anti-aircraft gunners and searchlight operators, we were able to find both aerodromes. We stooged around for a while without seeing any signs of activity, before dropping our bombs and turning for home. We certainly hoped we had been able to dissuade some of the bombers from taking off to deliver their bombs over our country.'

An extract from *Flüuge gegen England* [*Flights against England*], published by the High Command of the German Armed Forces (the OKW), 1941 said 'Quietly our commanding officer gave the pilot his approach instructions: 'A little more to the right, still a little more. Fine, now we're in position.' We came nearer and nearer. The horrible yet beautiful scene was close enough to touch. Dense smoke rose over the roofs of the city and stretched far into the countryside. We could distinctly see tall flames flickering upward. One particularly large fire among the countless others, showed where an extensive industrial plant must have been badly hit. We were over our target. The anti-aircraft artillery fired desperately. We were surrounded by the flashes of exploding shells. We could see clearly the vast extent of the fire-centres and the flames licking across large areas of this industrial town. At that moment our bombs too were released. A blow passed through the plane. Down below, the glow of fresh explosions shot up with the brightness of daylight. We were the first machine of a group of German bombers. Others had been there before us and more followed - until dawn of the new day that would reveal the full extent of the Coventry disaster.'

Writing about the Coventry Blitz, H. C. Legg recalled: 'It is 18.00 hours and promises to be a nice evening with a good moon. We have received a signal from Flight Headquarters to proceed to Gentlemen's Green to open a new balloon site. Everything was going according to plan. The high silver balloon was spread over the Green and we were just making preparations for its inflation when Jerry decided that it was just the night to send a few hundred bombers over and give us a taste of the things we had to come. The usual signals had been received; the Purple and Red and then sirens and finally the peculiar double drone of Jerry's engines. The order is: 'Take cover.' For some moments a queer sort of hush had fallen everywhere and now, looking back, all of us in that little crew of a dozen chaps knew that this was going to be the 'night of all nights.'

'We all stood looking into the sky and. with the exception of machinery in the distant factory and a shout or two from the wardens ordering some offender to take shelter; you felt you could have heard a pin drop. The droning above becomes louder and suddenly there is an ominous swishing noise that breaks into a scream and ends with a mighty crash. The show is on. Ack-Ack are firing all around and in a matter of minutes the whole scene has changed as bomb after bomb is crashing down in such numbers that it is absolutely unbelievable. Searchlights are creeping all over the heavens - Verey lights and flares dropped by Jerry that seem suspended in the sky. Fires are breaking out everywhere and still they come - Crash! Crash! Crash! The ack-ack are still blazing away and by now we are shouting to each other, being deafened by the roar. The light is terrifically bright and the scene all around is like a gigantic stage setting. The fields have turned a brilliant green and cattle can be seen racing aimlessly about. Fire engines - ambulances - cars - vans, all are rushing about, shouting and orders can be heard from all directions. On and on it goes, these flying lunatics seem to have no limits. Suddenly you realize that the searchlights have disappeared, it must be that they have no further use. A few moments later Jerry planted one as close to us as we wished. Fresh flames broke out and a few more homes were about to perish; and in some cases the occupiers as well - poor devils!

'This seemed the moment when even King's Regulations must go by the board and twelve able bodies went tearing across the Green on to the roadway, having not the slightest idea what they were going to do. Several fires were burning and many people were staggering about with shock, as though they were hopelessly drunk. Some folk were clearly visible in their shelters, the women trying to calm the children - some frightened and some almost bordering on madness. I could not think what was the best thing to do in such a horrible, hopeless situation and found myself going from shelter to shelter asking everyone if they were OK and telling them to keep calm as everything was all right. Several little duties we were able to perform, getting them more blankets and things that they needed which had been forgotten during their rush down to the shelters. More bombs had dropped close by and still that droning went on and on. Wave after wave was coming in and the sky seemed packed with Goering's planes. Bang! Crash! And so it had been for hours now - suddenly I realized there seemed to be no gunfire. My God! We had used all our ammunition! Now the aircraft were coming in much lower. Nothing could stop them. No guns, balloons or searchlights and a target as bright as an illuminated area. Out in the street again, a warden is looking up at the sky and mutters: 'Bastards.' I have to agree with him.

'One of my mates is minus hat and tunic. He had used the tunic to cover some poor blighter over and hasn't got a clue where his hat is. Suddenly from close by there are some heart-rending screams and you know that somebody has found the body of a relative or a sweetheart and still you are quite helpless. We make our way across the rubble and hear a friendly voice calling from somewhere: 'Hey, Chum, hey!'Ere you are, down 'ere.' We both look around and see a chap lying on his back close to a brick wall. He can't move his leg and says he is feeling cold. I take off my greatcoat and sling it over him. We take a door that is lying around to use as a stretcher. He laughs and says: 'Bloody good idea' and hopes he can play football on Saturday. When we move him he screams with agony, so we call a couple of wardens to help and we manage to get him on the door. He wants a smoke and tugs at it gratefully. A warden cuts his trouser and his leg falls completely apart as though cut with an axe - from thigh to ankle. He sees it happen and passes out. We carry him over to the First Aid Post. Next day my greatcoat came back with a pound note and a message in the pocket saying: 'Thanks, pal, have a drink with me.' The following day he died after they took the remains of his leg off.

'As we were making our way back a little group had collected outside a house that was on fire upstairs. Somebody was up there. We dash up - the heat is terrific and the situation is hopeless. Somebody shouts: 'Look out!' and I fall headlong down the stairs, my pal after me - and out into the street again. There is a wrenching of burning beams and the roof collapses. Flames leap higher, the heat is greater. The little group is standing farther away and now they have a tragic look about them. I wondered who had been in that house that was already only a burning shell.

'Still it goes on, this great crescendo of exploding bombs and now there are large open spaces where, a little while ago, had been blocks of buildings. One knew not whether it was night or day and could not care less. You were only praying for it to cease. It did not seem possible that one could endure this raging madness till the following morning; but it was to be and we staggered back to our site, blackened, tired and hungry.

'The following evening we walked over to the working men's club. As we approached we could hear them singing. We entered and they stopped and gave us a little cheer and brought us drinks. There was no electric light, only candles, a great tarpaulin over part of the roof. You could see some women's eyes were red with weeping and the men were grimy and tired. Some had been to their jobs as usual, others were having one for the road before going on duty and they sang songs like *Rose of England*. This was the first time in my life that I had encountered sheer guts and courage and it almost made one choke.' [27]

The raid that began on the evening of 14 November 1940 was the most severe to hit Coventry during the war. It was carried out by 515 German bombers, from Luftflotte 3 and from the pathfinders of Kampfgruppe 100. The attack, code-named Operation Mondscheinsonate ('Moonlight Sonata'), was intended to destroy Coventry's factories and industrial infrastructure, although it was clear that damage to the rest of the city, including monuments and residential areas, would be considerable. The initial wave of 13 specially modified Heinkel He 111 aircraft of Kampfgruppe 100, were equipped with X-Gerät navigational devices, accurately dropped marker flares at 19:20. The British and the Germans were fighting the Battle of the Beams and on this night the British failed to disrupt the X-Gerät signals. The first wave of follow-up bombers dropped high explosive bombs, knocking out the utilities (the water supply, electricity network and gas mains) and cratering the roads, making it difficult for the fire engines to reach fires started by the follow-up waves of bombers. The follow-up waves dropped a combination of high explosive and incendiary bombs. There were two types of incendiary bomb: those made of magnesium and those made of petroleum. The high explosive bombs and the larger air-mines were not only designed to hamper the Coventry fire brigade, they were also intended to damage roofs, making it easier for the incendiary bombs to fall into buildings and ignite them. At around 20:00 Coventry Cathedral (dedicated to Saint Michael) was set on fire for the first time. The volunteer fire-fighters managed to put out the first fire but other direct hits followed and soon new fires in the cathedral, accelerated by firestorm, were out of control. During the same period, fires were started in nearly every street in the city centre. A direct hit on the fire brigade headquarters disrupted the fire service's command and control, making it difficult to send fire fighters to the most dangerous blazes first. As the Germans had intended, the water mains were damaged by high explosives; there was not enough water available to tackle many of the fires. The raid reached its climax around midnight with the final all clear sounding at 0615 on the morning of 15 November. Coventry's air defences consisted of twenty four 3.7 inch AA guns and twelve 40mm Bofors. Over 6,700 rounds were fired. However, only one German bomber was shot down. In one night, more than

4,000 homes in Coventry were destroyed and around two-thirds of the city's buildings were damaged. The raid was heavily concentrated on the city centre, most of which was destroyed. Two hospitals, two churches and a police station were also among the damaged buildings. Around one third of the city's factories were completely destroyed or severely damaged, another third were badly damaged and the rest suffered slight damage. Among the destroyed factories were the main Daimler factory, the Humber Hillman factory, the Alfred Herbert Ltd machine tool works, nine aircraft factories and two naval ordnance stores. However the effects on war production were only temporary, as much essential war production had already been moved to 'shadow factories' on the city outskirts. Also, many of the damaged factories were quickly repaired and had recovered to full production within a few months.

An estimated 568 people were killed in the raid (the exact figure was never precisely confirmed) with another 863 badly injured and 393 sustaining lesser injuries. Given the intensity of the raid, casualties were limited by the fact that a large number of Coventrians 'trekked' out of the city at night to sleep in nearby towns or villages following the earlier air raids. Also people who took to air raid shelters suffered very little death or injury. Out of 79 public air raid shelters holding 33,000 people, very few had been destroyed. The raid reached such a new level of destruction that Joseph Goebbels later used the term Coventriert ('Coventrated') when describing similar levels of destruction of other enemy towns. During the raid, the Germans dropped about 500 tonnes of high explosives, including 50 parachute air-mines, of which 20 were incendiary petroleum mines and 36,000 incendiary bombs.

The American Associated Press News Agency reported: German bombers have destroyed the heart of this formerly peaceful town in the English Midlands in a raid that lasted from the evening dusk until dawn, turned parts of the city into an inferno and left at least 1,000 people dead and injured. Coventry's wonderful and famous red-brown sandstone cathedral is a pile of smoking ruins... Throughout the night the narrow streets through which Lady Godiva rode her horse almost one thousand years ago, trembled and cracked under the thunder of the dive bombers, the howling and explosions of the bombs and the crackling of the antiaircraft artillery.'

The attack on Coventry seemed to Jim Moore 'to bring about a change of policy for the bombers of the RAF, who, up to that point, had generally been instructed to aim for targets of industrial or military importance. In our case, at our briefing on 16 November, we were initially briefed to bomb the docks at Hamburg. Later, we were recalled to the briefing room to be directed to drop our bombs on the city itself. War in the air is impersonal but in the case of the navigator Sergeant Bob Weston, who was from Coventry and others from Coventry, or other English towns which had been bombed, they may well have felt differently as their families and friends were likely to be killed. In their case could you blame them if they felt this was an act of retaliation although, to be fair to Bob he never spoke about it in this way. At 1905 hours we took off on this operation, which was to last for four hours 25 minutes, against a target which was particularly well defended. The city being on an estuary was relatively easy to find, although the reception we received by way of flak and searchlights was pretty impressive. Bob picked up the target on to which he directed Roger, the seconds ticking away like hours as we flew straight on the bomb run, before these magic words, 'Bombs Gone!' and we were able to take some evasive action. Following this operation we were granted leave.'

On the nights of 15/16 November and 16/17 November some bombers were sent to attack aerodromes which the Germans were using for attacks on Britain, but all the rest - though this did not amount to what would later be considered a strong force - were put on Hamburg. And one at least of the targets in Hamburg was a large industrial area, the Veddel and Peute area.

On 4/5 December 83 aircraft set out for Düsseldorf and Turin but only thirty of the bombers reached these targets. A Flying Officer described the raid on Turin later in a BBC broadcast.

'This was the first time our squadron had done the Italian trip. We'd heard a rumour about a week before that we might be getting the job and everyone was quite thrilled at the idea of the run over the Alps. We were told in the morning that we were going to Turin and so we started at once drawing our tracks and getting the navigation generally weighed up. My navigator was particularly keen on the show because he's something of a mountaineer and has done a fair bit of climbing in the Alps. The route we were taking worked out at between twelve and thirteen hundred miles there and back. We had to make a bit of a detour to keep clear of Switzerland because we had special instructions to avoid infringing Swiss neutrality. Briefing was at two o'clock in the afternoon and we took off just as it was getting dark. To start with, the weather was poor and we had to come down to six hundred feet over the English coast to pinpoint ourselves, then we climbed up through what was becoming really nasty weather and crossed the coast on the other side fairly high. By that time the cloud was what we call ten-tenths - that's to say, it obscured everything, but eventually we got above cloud and then we had the light of the moon which was in its first quarter. Before that it had been very dark indeed. We were flying blind above cloud until we arrived forty or fifty miles east of Paris and then we ran into clearer weather, the clouds gradually decreased below us until we could see the ground and when we reached southern France the weather was perfect. It was one of those clear moonlight nights when the stars seem to stand out in the sky and you feel you can put out your hand and grab one.

'As we flew on towards the Alps, we could make out some of the little mountain villages against a background of snow, the whole scene resembled a picture on a Christmas card. The aircraft was going wonderfully well and we cleared the highest mountains we went over by three or four thousand feet. You could see the ridges and peaks, well defined and the moon shining on the snow was half turning the night into day. Flying over this sort of scenery was something completely new to us and pretty awe-inspiring. The nearest we'd got to it was on the Munich raid when we'd seen the Bavarian Alps in the distance. The navigator came up and pointed out Mont Blanc, away on our port side. He was able to identify it from its shape because he'd actually climbed it and he was telling us how he was beaten by the weather when he got to within six hundred feet of the summit. Immediately we got to the other side of the Alps, with no snow about, it seemed by comparison, intensely dark for a bit. It was like coming out of a lighted room into the black-out. Soon after that we started to glide down, losing height very gradually and arrived slightly west of Turin. Other planes were already over the target because we could see their flares and there was a barrage of anti-aircraft fire in the sky.

'Our target was the Fiat works and the whole time we were looking for them we were still gliding down to our bombing height. Actually we picked the works up in the light of somebody else's flare. They were unmistakable. I've never had such a target before. There seemed to be acres of factory buildings. We almost wept afterwards because we hadn't got any more bombs to give them. Having located our target we flew four or five miles away, turned round and made our run up over it. The wireless operator came along and stood beside me to have a look at the bombing, otherwise he wouldn't have seen anything from his usual position. He's a bit of a wag and when he saw the light flak coming up from the works he said: 'Gosh, look at the Roman candles.'

'We made two attacks and as we came round afterwards to have a look, the fires which we'd started were going strong. There was a big orange-coloured fire burning fiercely inside one block of buildings. Having finished the job, we climbed to get enough height to cross the Alps again. 'Altogether we were over or round about the town for three-quarters of an hour and whilst we were circling to gain height we saw somebody hit the Royal Arsenal good and proper. Going home, the Alps didn't look quite the same. The moon had almost set then and the mountains had lost their vivid whiteness. The last two hours of the journey home were, frankly, plain misery. It started with the aircraft suddenly beginning to get iced up. I tried to climb, but she wouldn't take it. Ice was coming off the airscrews and hitting the fuselage. We came down to about seven thousand to thaw out and then we ran into an electrical storm. All this time we were in cloud. It was frightfully bumpy and the aircraft was bucketing about all over the place. At one point the front gunner called me up and said: 'Are you quite sure you're flying the right side up, because I think I can see white horses in the sky.' That was when we were over the North Sea. When eventually we left the clouds we had to come through snow and sleet and the final bit of the journey we made in a howling gale which reduced our ground speed a lot. Never had we ever taken so long to get inland to our base from the coast, but we got there safely in the end.'

On 16/17 December in Operation 'Abigail Rachel' 134 aircraft set out to raid the industrial centre of Mannheim in retaliation for the German bombing of English cities, notably Coventry on 14/15 November and Southampton. The intention of Abigail was to cause the maximum possible destruction in a selected German town. The towns selected were Bremen ('Abigail-Jezebel'), Düsseldorf ('Abigail-Delilah') and Mannheim. The original selection was 'Delilah' and crews booked a date at 10.15 on the morning of the 16th only to have it changed three hours later after adverse weather reports brought a late change of plan and Mannheim was selected. Good weather was reasonably certain for the first part of the night but not for the early hours of the following morning. The aiming point was officially 1,500 yards south of the Motorenwerke-Mannheim, a works engaged in the production of diesel engines for U-boats but the moon was bright with good weather conditions and good visibility and all the later sorties had no difficulty in recognizing the town from the flak and the fires. Fourteen Wellingtons of 3 Group carrying their maximum load of 4lb incendiary bombs opened the attack, their fires providing a beacon for the following waves. 47 out of 61 Wellingtons, 33 out of 35 Whitleys, 20 out of 29 Hampdens and three out of nine Blenheims succeeded in finding and bombing Mannheim.

The story of one of the Wellington squadrons involved was as follows:

'After the regular morning air test the aircrews drifted back in ones and twos to their messes to discover the battle order for the night posted on the notice boards. Twelve crews were up-half the maximum effort of the squadron. 'M for Monkey', a Wellington bomber, was flying tonight. Her pilot and observer went into the mess together, noted their names on the battle order-and declined pre-lunch beers. They exchanged desultory squadron chat with the other officers. Already, an air of tension lay over the mess.

'The Senior Intelligence Officer, wearing 'Mutt and Geoff' ribbons from the First World War under his pilot's wings, came into the ante-room and said: 'Briefing's at three o'clock, chaps.' Almost at once his words were confirmed by the metallic voice of the Tannoy system blaring out over the airfield: 'All crews on the battle order are to report to the briefing room at fifteen hundred hours.'

'After lunch, followed by a nap in an easy chair, the two officers from M for Monkey walked slowly in the misty autumn sunlight to the long briefing room attached to one of the hangars. At the door an armed service policeman examined identity cards and cheerfully wished them good luck as they passed inside to where several rows of chairs stood facing a raised platform. Behind the platform a large map of northern and western Europe completely covered the wall. A white dust sheet hid most of Holland and Germany and a red ribbon, stuck through with large-headed map pins, led from the Lincolnshire airfield across the North Sea to disappear behind the sheet.

'The room was noisy with the babble of 72 men as they cloaked their nervousness with catcalls and banter. Suddenly it was hushed as flood-lights illuminated the stage and a group of officers, including the station and squadron commanders,' mounted it. Amid the preceding hubbub nobody had mentioned - or even in jest hazarded a guess - as to where the target for tonight would be.

'The squadron Commanding Officer, a young Wing Commander with a DFC ribbon gleaming beneath his brevet, stepped up to the map and watched with bated breath by the crews, ceremoniously pulled off the white sheet. There was a silence, a pause and then a deep sigh that was almost a groan as the crews saw that tonight's target lay deep in enemy territory.

'The Wing Commander spoke. 'OK, chaps, Mannheim again. Now you know the worst - it's a hell of a way but the flak there's not too bad. At least it wasn't, the last time I went. The intelligence officer will put you in the picture about aiming points, flak, etc.'

'A young intelligence officer stepped forward, a pointer in his hand. 'Yes, sir - well, chaps, Mannheim is a port on the Rhine, population.' The intelligence officer's voice sounded high and clear in his effort to make himself sound superior - but he wasn't and he knew it: fortunately, the crews knew he knew it, too.

'His briefing finished, the intelligence officer stepped from the stage to be followed by the observer leader who explained the suggested route, the armaments officer who described the bomb load and the wireless officer who dealt with the various codes to be used during the night's operation. Then came the meteorological officer, who interspersed his slides of weather charts with others of scantily dressed ladies which were greeted with ribald cheers.

'A pep talk from the station commander, a Group Captain, concluded the briefing and, as the senior officers left the briefing room, the crews filed out into their changing room, where they leisurely changed into flying kit, collected their flying rations - chocolate, chewing gum and barley sugar - and moved into the high room next door to collect their parachutes. Left behind in the briefing room, the observers plotted their routes on Mercator charts. By 1700 hours all except the observers had piled into open 3-ton trucks and were shuttling out to their aircraft, dispersed around the airfield in the grey Lincolnshire dusk.

'M for Monkey' stood squat and black against the skyline waiting for her crew to mount the short metal ladder which disappeared into the nose of the fuselage. There were a few caustic exchanges with the ground staff - hanging around pretending to attend to odd bits of equipment but in reality there to wave 'Good Luck' to their own special aircrew - then the crew disappeared up the ladder and out of sight. The observer, his arrival delayed by the route planning, arrived after the others in a small van and, laden with a large green canvas satchel and a sextant, was the last of the crew to enter the 'Wimpy'.

'Two members of the ground staff at a thumbs-up sign from the pilot slowly rotated the large airscrews and stood back while the pilot pressed the port-engine starter. A high-pitched whine, a puff of vapour, a splutter and the big engine started, making surprisingly little noise as it did so. The ground crew wheeled the starter battery on its trolley round to the other side of the plane where the same routine was run through with the starboard engine.

'While the pilot and second pilot completed their cockpit check, checking and cross-checking each item on the long list together, the observer pinned his chart to the green-painted navigator's table. He carefully adjusted the light over the table and placed his various instruments in handy places, wedging them behind wires or into joints in the cockpit furniture. Above his head, tied to the header tank of hydraulic fluid, he hung a small teddy-bear - the crew's mascot.

'Satisfied that all was well with the aircraft, the pilot switched on his intercom and called up each member of the crew in turn, checking that they were ready to go. When they had all replied he called up the control tower over the radio telephone link and informed the Duty Officer that 'M for

Monkey' was ready for take-off. An answering green flash gleamed momentarily from the control tower and the pilot released the brakes. Slowly 'M for Monkey' began her journey to Mannheim.

'As she taxied over the grass to where a small van showed the take-off point, the crew strapped themselves in and made final preparations for the journey. A small knot of men standing around the van gave the thumbs-up sign and the aircraft lined itself up with the row of flickering lights. The pilot switched on his intercom, as did the second pilot. A green Very light soared up from the top of the van, the pilot answered by a flick of the landing light and 'M for Monkey' had been cleared for take-off. Inside the aircraft the crew sat tense as the pilot pushed forward both throttles. The fuselage shook and the wings seemed to undulate as he released the brakes and the heavily laden aircraft began rolling slowly over the grass. As it gathered speed the second pilot read off figures as the needle climbed and when the pilot was satisfied, he hauled back on the control column and the Wellington lumbered slowly into the air. Safely airborne and the first course set from over the middle of the airfield, the crew busied themselves with their tasks. The gunners, front and rear, began their long stare into the black night sky, acclimatising their eyes to its emptiness; the wireless operator, at his table behind the observer, slowly twiddled the knobs of his various sets so that they would be set up on the correct ground stations for an emergency call. The observer, illuminated only by the light reflected off his navigation table, was comparing the course drawn across his Mercator chart with unfolded topographical maps spread over his knee.

'Side by side in the cockpit the two pilots sat almost motionless as if they were asleep, their heads hunched into the upturned collars of their thick fur-lined flying jackets. Occasionally one or other of them would come to life and point to a dial in front of him, to be answered by a thumbs-up sign or by some alteration to one or other of the many control knobs. After about 20 minutes on their first course, the second pilot broke the long silence on the intercom to tell the observer that he had caught sight of the coastline ahead of them. The observer came out into the cockpit and stood behind the second pilot, searching for an accurate pinpoint from which to start a dead-reckoning plot. As the pilot could fly whichever route to the target he wished, an accurate air plot was essential in case he decided, for any reason, to alter the route. Satisfied at last, the observer went back to his table and plotted the pinpoint before calling up his captain to give him the amended course across the North Sea to the point near Ostend where they would turn east. Flying this latter course until the Rhine was reached south of Koblenz he would then map-read their way south to the target area.

'Once well away from land the rear gunner, his voice slightly breathless as a result of the cold rarefied air, asked permission to test his guns by firing them into the sea below. The pilot gave his permission and then, his memory jogged by the sound of the rear gunner's voice, told the second pilot to switch on the oxygen. At the same time he checked with the observer that the IFF (Identification - Friend or Foe) was switched off. The crew, even behind their oxygen masks, could smell the cordite from the gun test and were reminded of the dangers of their mission. From that moment on a new height of tension pervaded their actions.

'As they passed over Ostend, the land beneath them was covered with low cloud and the plane had to set course for Koblenz without a pinpoint. When, two minutes later, unseen by the observer in the blacked-out cabin, the sky beneath them was filled with a tracery of red, green and yellow light flak, the pilot switched on his intercom and commented, 'That's Ostend, I reckon - unless it's Ramsgate'. The observer, amused by neither the lack of a pinpoint nor the heavy joke about his navigational prowess, began to haul out the several volumes of almanacs and tables necessary to plot an astro-fix.

'M for Monkey' forged on, eastward now, over the dark hills of occupied Belgium. Ninety minutes later, the observer plotted his second astro-fix, gained by sextant observation from the moon and two stars. He switched on his intercom, gave the pilot permission to continue the weaving flight he always liked to perform after the aircraft had crossed the enemy coast and said, 'We're on track OK - I'm coming up in a minute to see if I can get a pinpoint. If you see anything like a river, sing out.' Picking up the map of the Strasbourg area he unplugged himself from the intercom circuit and the oxygen supply and went through the cabin door to stand behind the second pilot.

'Below the aircraft the cloud had now disappeared and the landscape of wooded hills showed up clearly in the bright moonlight. All the crew - except for the wireless operator, happily tuned into the BBC Home Service - were staring into the night, searching for the glitter of water. Suddenly the second pilot pointed downwards to starboard. The observer said, 'That's the Nahe, the Moselle, or the Rhine. Let's go and have a look and I'll see if I can fit it on to the map.'

'Later, his position satisfactorily fixed, the observer went down into the nose of the aircraft and assumed his second identity as bomb-aimer. Lying on his stomach, a closely-shaded orange light faintly illuminating his map, he map-read the course until the Rhine was directly beneath them. At the same time, working in the dark almost as if by instinct, he began to prepare the bombing panel and bombsight for the bombing run and attack. Telling the pilot to turn south-east and to track along the right bank of the river, he concentrated more and more intently on the ground beneath him.

'The south-easterly course took the aircraft over Bingen and now Mannheim lay only ten minutes ahead of them; the Rhine flowed away to the left, the broad river growing narrower in the distance but still reflecting the moonlight. No one spoke and, apart from the heavy breathing of the rear-gunner who kept his intercom 'live' so that he could give immediate warning of any enemy fighter, each member of the crew silently faced the stress and strain of the next ten minutes.

'Suddenly the sky ahead of them, until then a dark and velvety blanket, was punctured with red flak 'bursts. 'Mannheim!' grunted the pilot over the intercom and the observer silently prayed he was right; after a previous raid on the city, reconnaissance reports had shown that most of the bombs had fallen in fields near Heidelberg and that the railway yards they and the other crews were supposed to have attacked had escaped unscathed. Still in the distance, over to port, the Rhine gleamed, a welcome reassurance that this time their navigation was correct. Quite suddenly the river appeared to swing towards the aircraft and the observer hurriedly checked the compass in the bombsight to make sure that the plane itself had not altered course towards the river; but just then a large town swung beneath him. 'Worms,' he thought, doubly sure because of the odd-shaped lake south-east of the town. He was now convinced that they were heading for Mannheim with about only 10 miles to go - especially as ahead of them the sky was now full of flak bursts.

'Telling the pilot to continue on course until he saw the river divide, the observer looked along the wires of the bombsight. Suddenly the junction of the river appeared - 'I've got it!' yelled the observer. The pilot replied, 'I can see it too, but I'll go round a couple of times to make sure it really is the right place this time! Keep a good look out for Jerry and watch we don't get too close to any of our bods either.'

'As the aircraft swept across the target area white flashes that pocked the darkness beneath them indicated that other aircraft were dropping their bombs. All around flak bursts lit up the sky, in the distance red, but yellow and white when close. Springing up from the ground below, an occasional searchlight waved its silvery finger of light.

'Twice the pilot took the aircraft through two slow turns round the aiming point, before flying away to get the moon behind the target and help the observer to identify it more easily. Approximately lined up on the target, the pilot told the crew that he was about to start the bombing run. The tension by now had increased almost to breaking point, but the pilot eased it slightly when he told them that they could throw out their own personal contribution to the raid - empty beer bottles.

'The only voice on the intercom now was that of the observer. 'OK skip' - keep her where she is - left, left a bit more - blast that searchlight - left, left again - OK skip', steady-steady-steady - bomb doors open - right-right-steady, keep her there.' Suddenly his voice lifted excitedly as he pressed the bomb release button. 'Bombs gone!' he shouted. 'It looks good this time, skip, I could see the docks quite clearly.'

'Immediately every member of the crew appeared to be shouting over the intercom, giving their own personal views of the success of the raid. Cutting through their voices the pilot told them to be quiet, the authority in his voice calming them down. The observer, after a cheerful tap on the hanging foot of the front gunner struggled back through the cockpit to his navigation table and plugged himself back onto the intercom and oxygen.

'He asked the pilot what course they were flying.

'Two seven zero. Ludwigshafen right beneath us.'

'Satisfied, the observer bent over his log and chart and worked out his course for home.

'Now that the excitement and elation of finding and attacking their target had ceased, the pilot warned the crew to be particularly watchful for enemy fighters and at the same time he asked them - while memory was still recent - for their views on the attack. The observer recorded the crew's comments in his log.

'The rear-gunner had seen several bombs exploding in the barge dock area as they flew away from the target and the second pilot reported that he had seen another Wellington crossing only a few hundred feet above their own aircraft, its silhouette sharp in the bright moonlight. Other reports were confused but optimistic.

'The red flak bursts behind them gradually disappeared and the crew turned their minds to home and the promise of clean sheets and warm beds. They thought of the scene at interrogation, when over cups of Ovaltine laced with navy rum they would tell the intelligence officer of their success. He, piecing together the reports of all the squadron's crews, would make his own estimate of the success of the raid and send it on to Base HQ.

'But only the Germans would know for certain if 'M for Monkey' and her companions had struck home.'

Eighty-nine tons of HE and nearly 14,000 incendiaries, including a number of special 250-pounders were dropped in the six hours of attack. Bombs were clearly seen to fall all over the target area and countless fires and many large explosions were referred to in all reports. In the southern part of the target area dense black smoke was observed as though from an oil fire. Aircraft arriving late in the night reported that many blocks of buildings in the western and south-eastern areas of the target were ablaze and at Neckerstadt there was a continuous series of explosions thought to be from a munitions dump. Losses in aircraft and crews were slight. A Wellington crashed on take-off killing three members of the crew. A Hampden and a Blenheim were lost without trace while a second Hampden crashed in the English Channel with the loss of the crew. Four more aircraft crashed in England on return, at Rye and Hastings in Sussex and at Plympton and Norwich and a Wellington overshot the runway at Mildenhall. Pilot Officer Brant on 10 Squadron brought his Whitley back from south-west Germany to Bircham Newton on one engine, jettisoning guns, ammunition and all loose objects to maintain a height of 2,000 feet over the sea.

Here is a German account of that night in Mannheim:

'What we poor people have to endure here: terror, fire, fear! On 16/17 December we experienced a great attack - words cannot describe it. Seven hours in the cellar - there were said to have been more than 200 planes; a rain of bombs over our town, fires everywhere-no light, no water. We were not even dressed when bombs began to fall and at once the lights went out. There was a terrible blaze. The greatest shock we had was the burning down of the oil mill and the danger to life. I cannot tell you what we went through. People were in their cellars all night - the houses burning above, nothing but incendiary bombs. There was no Fire Brigade, no water, so everything went on burning like the oil mill; the Karlsruhe Fire Brigade [Karlsruhe is over thirty miles due south of Mannheim] finally put the fire out. The next day the bombs were exploded - our house shook - we dread the night and go to bed with our clothes on; it is so terribly cold and we have so little coal. It causes a lot of deaths more than one knows. If it goes on like this the people will be finished.'

Shortly after the raids on Mannheim a Squadron Leader DFC recounted the operation on the BBC. 'The operation' he said 'was on a pretty big scale; aircraft from a number of squadrons were operating. The general idea was to send in the early ones with incendiaries so as to light up the target, then for the main force to come along with heavy stuff. The operations of the main force incidentally were spread over a period of six or seven hours.

'The station commander had a word with the captains and crews before we left. He said he was expecting some very good results. He also mentioned that, as there would be a lot of aircraft concentrating on Mannheim, he wanted us to go in, bomb and come out again as quickly as possible.

'We left at regular intervals. It was important to keep strictly to the scheduled take-off times because of working in with the other stations, so as to make the bombing a more or less non-stop affair once it started. Just when we were due to get away it started raining cats and dogs. One could just see a few blurs of light indicating the flare-path and that was all - rather like driving a car in heavy rain without a windscreen wiper, only more so, if I may put it that way. However, we all got off all right. The cloud base was at a thousand feet and we had to climb up to get through it. We were climbing rather slowly, too, because we were carrying a heavy load. Once we got above the clouds, we were in bright moonlight and the navigator got his sextant out and started taking Astro sights to check up on our position. Normally, when you cross the Dutch coast you reckon to get a bit of flak thrown up at you but this time, being still above cloud, we got nothing at all. We flew on, keeping straight and level. Then, fifty miles inside the Dutch coast, the cloud cleared and we saw the ground for the first time since we'd taken off. Altogether, it was a very uneventful trip out. In Germany they'd had a fall of snow which was quite a help to navigation. When you have a light fall, as this was, the important things - woods and rivers, lakes and towns and villages - all stand out much clearer and so, with the moon very bright, we pin-pointed ourselves quite easily as we went along.

'We were some distance from Mannheim when the front-gunner reported heavy flak ahead. We were then about ten minutes away heading straight for it and we knew it must be Mannheim. The stuff was coming up in bursts and then dying away, then breaking up again, spasmodically. I told the navigator to prepare for bombing and he came up into the bomb aimer's position in the nose of the aircraft with his map. Having done that, he had to check up on the bomb switches, select his bombs and we determined the length of the stick. One can drop a widely spaced stick or a close one. This time I had decided on a very close one.

'As we approached, I could see fires already well under way and it was obvious that the blitz was in full swing. We picked up the Rhine, followed the river up and then started to take avoiding action because there was quite a lot of flak, mostly light stuff, coming up. It's all tracer, this light stuff and you can see strings of it coming up.

'The flak seemed to be pretty continuous by now. When it gets like that, one just goes through it, doing evasive stuff. I don't think flak deters any of the fellows from carrying out the job. One sometimes sees German reports of the barrage turning our aircraft back. In my opinion, that's just nonsense. As we got a bit closer, the navigator called out 'Ready' and I levelled out and opened the bomb doors. You only do that at the last minute because when they are open it makes the aircraft drag a bit, so you open the throttles a little to compensate the slight loss of speed.

'You tell the navigator 'Bomb doors open, master-switch on' and he repeats that back to you. He will probably make a few corrections to course-'left, left, right, right, steady' and so on - and when he's bombed he calls out 'Bombs off'.

'As a matter of fact you can feel the bombs go. You get a slight lift in the aircraft and it immediately becomes much more lively. I must say, I always find it a bit of a relief directly they've gone and one knows that one's done the job one was sent out to do. Well, on this occasion everything went normally and as soon as the bomb aimer said 'OK sir, bombs burst,' I put the aircraft into a steep turn to let the crew have a look. There were three groups of huge red fires burning down below and spirals of heavy black smoke rising above the town. The fires were increasing in intensity all the time. Then, we set course for home, we could see other people bombing as we came away. I told my rear gunner, as I always do, to note the time when he could no longer see the fires and we were about sixty miles away when he called out and said he'd lost them. Nothing very much happened on the way back. Coming down through the clouds over the North Sea we started to get iced up a bit. Pieces of ice were threshing off the air-screw and coming through the fabric of the fuselage, but we got back without any real trouble. Everybody had seen much the same sort of thing as we had-fires and still more fires and the general feeling in the squadron was that the show had been a great success.'

Though results reported by all successful crews were 'satisfactory', on 21 December a PR Spitfire photographed Mannheim in daylight and the photographs clearly showed 'a wide dispersal' of the attack. From the mosaic, it was immediately apparent to Sir Richard Peirse the C-in-C that the operation had 'failed in its primary object'. Most of the damage occurred in the residential area of Mannheim but several bombs fell across the river at Ludwigshafen. Many aircraft were reported to have bombed from a great height without aiming, owing to the strength of the anti-aircraft defences. Sixteen large and 75 medium and small fires were caused, fire-fighting being made difficult by bomb damage to the main water system and by the freezing of water brought from the Rhine. A sugar and a refrigerator factory were put out of action, the Mannheim-Rheinau power station was damaged and output of armoured fighting vehicles and tank components from the Lanz works was cut by a quarter. Four other industrial plants were also hit. Twenty-three Germans were killed and 80 injured.

A small force of bombers went to Mannheim the next night to prolong the disorganization of the city. Before they reached Mannheim the crews saw fires still burning which had been lit the night before. From reconnaissance and from other sources it was learned there was great damage to railways, to shipping yards - Mannheim is a large river-port-and to factories; the chemical works in the suburb of Ludwigshafen; which faces Mannheim across the

Rhine, were badly damaged. It is evident that the change of policy was fully justified.

Such heavy and concentrated attacks could only be made when the weather and other circumstances were favourable. But whenever possible throughout the winter the policy was maintained of concentrating each night on one target, or at most on a very few targets. Only very occasionally was there a return to something like the scattered attacks of the summer of 1940; this was when our bombers were attacking targets too distant for a large force to reach at that time. Thus on the night of 12/13 January attacks were made on Porto Marghera, just opposite Venice on the Adriatic coast, on factories at Turin and on Regensburg, or Ratisbon, a petroleum port on the Danube; that night the port of Brest, a number of invasion ports and aerodromes in the invaded countries were also attacked. But this was an exception which proved the rule. It would be unwise to send a large force to very distant objectives, because a considerable proportion of the force might waste their effort, since the greater the distance the greater the chance of going astray. In this way the conditions of the summer of 1940 were reproduced and with them the mode of attack, the bombing of a number of small but vital targets by comparatively small numbers of bombers probably manned by particularly experienced crews. The leader of the squadron which went to Porto Marghera - this was 730 miles from base - was a most daring and resourceful wing commander. Wing Commander (later Group Captain) 'Speedy' Powell OBE DSO called 149 Squadron together at two-thirty in the afternoon before the attack and told them what to do. The target, he said, was a vital one and 'it is going to be a terrible waste of effort if we fly all that distance and then don't make absolutely certain of getting it.' [28] This meant that the pilots must come as low as they dared - not because the anti-aircraft fire would amount to anything much, but because the lower they came down the more they would have to climb to cross the Alps and the more petrol they would use out of a load that was just enough for so long a flight.

When Sergeant Jon Rootes on Flying Officer, the Right Honourable Brian Grimston's crew and others on 149 Squadron assembled in the Operations Room on the afternoon of the 12th January for a briefing on the planned night's raid, all eyes went, as usual, to the big map of the British Isles and Europe. The red ribbon was pinned at one end to our base at Mildenhall, Suffolk and the other end, stretching across Europe, the Alps and down through Northern Italy, was pinned to the target, Venice Porto Marghera. A round trip of some 1,500 miles, most of it by dead reckoning (pencil, ruler and thumb) as there was very little in the way of navigational help. The Squadron had flown two previous raids to Northern Italy, to Turin on 4 December and on 18 December 1940. This was the biggest one yet. 'During the briefing we were told that six aircraft would be on the raid; led by Wing Commander 'Speedy' Powell and that my aircraft - 'O-Olive' - would carry a new flask bomb camera in the bomb-bay to record where the bombs landed. Our bomb-load would consist only of three 500lb bombs instead of the usual eight, as the rest of the weight was in fuel contained in over-load tanks which would be jettisoned after use.

To Sergeant Rootes it seemed a long way to go just to drop three bombs! However, later reports confirmed that damage to the morale of the civilian population was severe and the panic caused in Italy by the raid was overwhelming. Take-off was scheduled for 18.00 hours but by then the airfield was blanketed in dense fog and the raid seemed in danger of cancellation. The Met men said the fog extended to 300 feet, with clear weather above that height. By 1930 hours the fog had thinned and Wing Commander Powell decided they should go. The Met men were correct and the aircraft popped out of the fog into a bright, clear, but moonless sky. Over France he saw that the ground was covered in snow. At 14,000 feet the temperature had fallen to minus 30 degrees. Over Lake Geneva the Wellington's engines coughed and stuttered and five voices, in unison, shouted to the second pilot, who was resting on the bed, to switch over the overload fuel tank. With the new supply of fuel the engines picked up again. Although Switzerland was neutral, Bomber Command's aircraft often overflew it to save on fuel.

The Swiss, to preserve their neutrality, would fire a small gun that reached to 5,000 feet. As the group approached the foothills of the Alps at 15,000 feet the moon revealed the cruel mountain peaks standing up like giant fingers as if seeking to claw the aeroplanes down. Mont Blanc to their right appeared higher than the aircraft. The temperature was now minus 50 degrees. Sergeant Rootes, without the luxury of turret heating, had long since lost all feeling from his waist down and was picking icicles off his chin as his breath froze. It was an unfriendly place to be and when the Wellington had cleared the mountains and begun to descend the crew voiced their relief. The navigator found Venice and could see its towers and palaces seemingly floating on the lagoon like a fleet of galleons.

Flying Officer Grimston banked 'O-Olive' to bomb the oil refinery at Porto Marghera scoring a direct hit in the middle of the tanks. A flash bomb went off at the same time with a brilliant white light and the camera took its photograph. Then Grimston dived the aircraft on the aerodrome at Padua and levelled out at 500 feet, giving the front and rear gunners some practice. The aircraft then turned for home, climbing back to 15,000 feet. [29]

The wing commander himself came down to 'ground-level' just north-east of the target. He flew past a fort where there were two sentries, who fired their rifles at the bomber as it went by; the bomber's gunners fired back. Then Powell flew on over the roof-tops of Mestre, seeing only a few people - 'ARP wardens or drunks, I expect' - in the streets. At Porto Marghera he had to climb to bomb; his own bombs would have blown him to pieces if he had not done so. Even so, he did not go high enough; he told his bomb-aimer, at 700 feet, to drop a small bomb, but the bomb-aimer dropped the largest he had; right on the oil refinery of Porto Marghera and this blew the aircraft violently upwards. The wing commander thought the bomber had been hit by flak - there were tracer bullets whizzing past in all directions - but all was well and he flew round again to use the rest of the bombs on the violent fire that was now blazing in the refinery. Then he came down to ground-level again and flew along the railway leading to Padua. He passed three goods trains: the driver of the first train waved, the driver of the second spat and the driver of the third looked anxious. He flew over Padua, dropped leaflets among the crowded towers and domes of that town and then, at twenty feet from the ground, his gunners machine-gunned the hangars of an aerodrome outside the city. The aerodrome's gunners opened fire and the wing commander made off, keeping as low as he dared and using every possible tree as cover. Then he climbed for the Alps. 'My tale,' the wing commander said, 'is the tale of all the rest of the squadron. They had all bombed from a low level; they had all scored direct hits; they had not wasted a single bomb. The long flight had been worthwhile.' [30]

Across the Alps the frost came down covering Grimston's Wellington in a white coat from end to end. The temperature plummeted and it was deathly cold inside the aircraft. Soon afterwards 'O-Olive' flew into thick cloud, the crew seeing nothing for three hours. At 0400 hours the navigator calculated that Calais should be beneath them and the Wellington started a descent. At 1,000 feet the cloud was still thick. At 400 feet, with the cloud almost touching the sea, the Wellington broke into a hazy light with the rain pouring down. They were over the Channel and ahead was the North Foreland light. Home was five miles away. The Wellington turned to starboard and flew up the Thames Estuary, avoiding the Harwich balloon barrage and let down into Mildenhall, ten hours after taking-off. One of the Wellingtons did not return. [31] A flare had caught light on the chute and the aircraft had crash-landed in the Venice lagoon. The crew had been picked up by the Italian police to spend four years in captivity.

The crew of 'O-Olive' went to their beds after debrief but were not to sleep in peace. Their Majesties King George and Queen Elizabeth were visiting Mildenhall and had been shown the photographs of the raid and wanted to meet the crew who had taken the shots. The sleepy crew were paraded in the Photographic Section and were questioned by the Royals who showed a remarkable knowledge of navigation, gunnery and wireless and the prime straight line of crew before her soon relaxed into a circle of friends. Outside it was snowing and a lone Spitfire flew around the airfield on guard. After the King and Queen had left the base, Sergeant Rootes retired once again, to snatch a few hours asleep before the next briefing at 1600 hours.

Later, in a BBC radio broadcast, a twenty-three-year-old Flight Lieutenant told listeners about the raid on Porto Marghera.

'That was the longest trip I had ever done. One knew it was going to be a pretty tiring business - ten hours or so there and back-on the other hand one was very pleased to be given the opportunity of doing something that was really new. Once before the squadron had been briefed for Italy and we'd been disappointed by having the operation cancelled because of bad weather. This time everything looked absolutely perfect from the word 'go'. It was just after six o'clock in the evening when we got off. The navigator pin-pointed himself on the coast, then we made a slight alteration of course and got our next pin-point on the other side, crossing the North Sea in comparative darkness because the moon hadn't risen yet.

'I decided we were a little too low to cross the coast safely in case we hit a heavily-defended area, so we turned left along it to gain an extra fifteen hundred feet. While we were gaining height in that way a fighter passed very close to our tail, but he evidently didn't see us. At any rate, he didn't attack. The rear-gunner, of course, wanted to have a crack at him - all air gunners always do - but it had come and gone too quickly for him to get his sights on it.

'We were taking as direct a route as possible to avoid passing over neutral territory. Seventy miles inland and ground became snow-covered. It was a fairly cold night - minus 20 degrees at 12,000 and later at 15,000, minus 25.

'We were climbing gradually all the time, until we'd reached a good height for crossing enemy territory: and we continued at that height. It was then just after nine o'clock. Half an hour before reaching the Alps we started to climb again and went up to 15,000. If it hadn't been such a good night from a weather point of view I should certainly have gone higher than that to get across. As it was, the winds were slight and there was no cloud, so there was no danger either of unpleasant bumps or icing conditions. We came to the foothills of the Alps after about three and three-quarter hours' flying. It was a nice clear night with everything in our favour. A lot of people burble about the beauties of the scene. Well, personally I must say, I was rather more impressed by the fact that if anything went wrong there was little chance of coming down safely, except perhaps by parachute. It was difficult to tell one peak from another. We were up at 12 or 13,000 feet.

'I suppose it must have taken about an hour from the commencement of the foothills on one side until we were clear of the foothills on the other side. The moon was still not up and having crossed the Alps, it wasn't at all easy to see any detail on the ground now that there was no snow to help us. In fact, it was so dark that we turned along the river, thinking for a moment that it was the coast. Eventually we came out some miles from Venice. The navigator recognized Venice from the form of its waterways and then we turned along the coast. Meanwhile the bomb-aimer was adjusting the various settings on the bomb-sight, such as, height, wind and air speed and trail angle. There was still no moon but a coast is always easy enough to follow on a clear night, however dark it is. We could see the outline of Venice very clearly defined as well as the famous lido and the bridge connecting Venice with the mainland.

'The target was a petroleum works which lay on the mainland just west of the end of a bridge, near the docks. Its position was quite obvious as we knew exactly where it was and we could have bombed it without a flare, but I decided it was quite worth while dropping a flare and having a look at the place before we attacked it. We ran up along the bridge, dropped the flare, did one circuit round it and then went out again and came in to make the bombing run from the same direction. The navigator watched the bomb burst on the target area and we did another circuit and saw parts of the plant going up in flames as the incendiaries dropped. There was a train crossing the bridge at the time and the explosion must have been a bit of a shock to the unsuspecting Italian passengers. Altogether, we' were about twenty minutes over the target area. Then we turned for home and as we approached the foothills of the Alps on the way back, the navigator who was in the astro hatch, said it looked as if the moon was sitting on top of a peak. We climbed to 15,000 again to get over the high part. The Alps looked a little more friendly now. That may have been due to the moon, but probably the fact that we were on the homeward journey had something to do with it too. Frankly, we were none of us sorry to see the last of the mountains.

'Just this side of the Alps we ran over cloud which blanketed out the ground. Then, having got clear of that, there was a lot of ground haze which made visibility very poor and again we were flying on dead reckoning navigation.

'On the English side of the Channel there were ten-tenths cloud at 4,000 feet and we were flying above it, in fact, we were not able to see the ground at all. I was pleasantly surprised, when we got the 'Over' signal - that's to say a signal from the ground, 'You are over the aerodrome.' However, my aircraft had behaved magnificently over the whole journey and had taken us there and back in the very good time of nine and a half hours.'

Pilot Officer A. A. Halley who started out on the night of 14 January 1941 to bomb aerodromes in Norway got into a gigantic cumulus cloud which towered up for about 16,000 feet, he saw a purplish glow develop on the right wing-tip and then it seemed that the wing-tip was on fire, but instead of the fire streaming back, as one would expect, it vibrated in front of the wing-tip. He dodged out of that cloud as quickly as possible and thereafter went round the massive cloud peaks or navigated between them so far as was possible. Nevertheless, he got caught in another disturbance which lit up the leading edges of the wings and the wing-tips in blue light, while the propellers became arcs of blue flame. Suddenly from the starboard wing-tip a great jagged spark of a bright purple colour shot out like forked lightning-it was huge, about thirty feet long. There was an explosion followed by a bright flash inside the cockpit and all the luminous dials and instruments stood out brighter than ever before. At the same time the wireless apparatus blew up and a bit just missed the face of the wireless operator, who was temporarily blinded by the flash. The bomber crew were lucky, inasmuch as they had only just received a wireless bearing from their base which enabled them to navigate safely back to Scotland.

Flight Lieutenant A. de Villiers Leach was caught in similar electrical storms on that night of January 14th. Flying at the two mile level to escape the lower cloud, he saw the cumulo-nimbus clouds reaching up and up to 20,000 feet, two miles above him. 'They were lit up by the moon and looked like giant peaks,' he said when he landed. 'They got higher and higher and surrounded me and it was getting so cold that I thought it better not to go over them and came down through a gap to 4,000 feet where we flew in cloud. A bluish flame two or three inches wide appeared at the tip of the propeller arc and gradually increased to about a foot in depth. When I looked out to port, I saw the whole wing fringed with the flame which looked like a bluish-white aura. Suddenly there was a terrific flash - I was absolutely blinded for a moment - and when I looked again the aura had vanished.'

Another pilot that night saw the whole of the metal surfaces of his wings covered with a bluish light and dancing tongues of fire, yet these young men who ride the storms and set the lightning flashing treat their terrifying experiences just as coolly as they stage their lighter interludes in the air.'

An assignment to bomb the railway junction at Rheims in the summer of 1940 gave Flying Officer Alexander Webster DFC*, a tall, dark-eyed Scottish Hampden pilot on 61 Squadron at Hemswell, a very uneasy moment, although it had its amusing side - afterwards. Rheims was duly located and the captain of the aircraft had what he called a 'beautiful horizon' as he went down to 2,000 feet. Then the searchlights caught them and the tracer started to come up. The pilot, thinking he would be clever, did a stall turn, but it did not work out as he intended. He closed the throttles as he put the nose up until the bomber slowed and stalled, then he kicked the rudder hard over while it turned to go down steeply. Unexpected things happened. He not only lost the searchlights, but also his beautiful horizon; all the instruments on the panel went completely mad. Sergeant Bisset, who a moment before had been shooting from the rear cockpit at the searchlights on the' ground, found himself on his back, with drums of ammunition falling around him, shooting at the clouds.

The captain, struggling with the control column while the altimeter touched zero, was staggered. He simply could not understand what had happened. A crash seemed certain when he regained control. At once the insistent voice of the navigator came to his ears. 'There's something wrong. It's oil - I can smell it, feel it and taste it.... You'd better get ready in case something happens.' There was a short pause as the navigator investigated in the dark. 'I think there's a bomb in front. I'm leaning against it!' he added. A minute or two later when they were safely over the sea and getting their breath, they switched on the lights to see what had happened. 'A yell came from the navigator. 'It's not a bomb - it's the lavatory!' he cried in amazement. It was and its normal position was in the back of the aircraft.

No magician ever performed a more difficult or dramatic conjuring trick on the stage than Flying Officer Webster performed that night in the air without any rehearsal. Before the remarkable feat could be accomplished, the Hampden had first to turn on its back and then dive for a while until the portable lavatory slid exactly opposite the well leading to the front cockpit. Only the most perfect timing could turn the Hampden right side up at that exact moment to enable the lavatory to drop into the well. A fraction of a second's delay would have sent it too far and made it fall on the head of the pilot. After turning right way up at that psychological moment, the Hampden had to dive again and do a bit of a climb.

Then, hey, presto! The 'bomb,' which happened to be neither unclean nor deadly, was transferred in a twinkling from the back of the aircraft to the back of the navigator in the front cockpit. The only explosion which took place was one of uproarious laughter from all concerned.

Steve Challen, a Wellington gunner on 40 Squadron recalled:

'Our first operational sortie scheduled for 27 December 1940, the target Le Havre, take-off 16.30. Just before we made our way out to T2515 we were issued with a flask and a bar of green foil-wrapped plain chocolate - the best ever issued! No escape aid package, Benzedrine pills, silk map and no flying-boots that could be separated at the ankle leaving the bottoms looking like civilian shoes. None of these helpful items for us. I had just the stuff that was issued me at Manston a year before; Sidcot brown-padded inner suit, long silk gloves, light leather gauntlets, black sheepskin calf-length flying-boots, helmet with earphones, oxygen mask with microphone and goggles. Over the top of the Sidcot came a parachute harness suit known as the Goon Suit which had the clips to attach the chest-type 'chute. What a job to get into the turret the first few times, especially if the 'chute had been clipped up to the upper starboard curved-in side of the turret. It was easier to get in, then lean back and bring the 'chute over the top of your bulk. All this exertion usually made for heavy breathing causing the perspex to steam up, more sweat trying to clear it.

'During the next two weeks we were issued with sheepskin flying-suits that had plugs for heated gloves and boots. I could only just get my Goon Suit over the top and the zips under the chin were hard to start. Wilhelmshaven, 16 January 1941, was a tryout for the gloves and boots, which seemed to get too hot. The only way to control the heat on the extremity, which was too hot, was to unplug. Not an easy thing to do, especially reaching your ankles where the connections for your boots were located. Another most important extremity was very hard to reach when you wanted to pee in that empty bottle! What a struggle, sheepskin trousered flap, parachute harness straps in the way and the centrepiece of the Goon Suit (about three to four inches wide) which came between the legs from the back up to the front under your chin where the zips connected: zipping downward making a tight fit around your legs. Not a situation for urgency! Removing your gloves to use the bottle, your hands soon became cold enough not to know when you returned them into the heated gloves whether they were hot, warm or cold. I suffered blisters on occasions through not feeling how hot the gloves had become.' [32]

Endnotes Chapter 2

22 *Fire Over London: The Story of the London Fire Service 1940-41.*
23   On 21 November the Germans announced that they had 'plastered Birmingham, the centre of the British armament and supply industries, with bombs'. According to the enemy High Command, 'in a succession of attacks hundreds of bombers had discharged more than 500,000 kilogrammes of bombs (nearly 450 tons), some of the heaviest calibre'. Fires and explosions were visible at a great distance and they 'were even more widespread than those in the raid on Coventry'. At least 682 Brummies were killed during the raids of 19-22 November. Another 1,087 were injured seriously and many more were slightly hurt. Heavy raids continued throughout December 1940, one of which lasted for 13 hours. Thence there was a long lull, interrupted by three raids, until the night of 9 April 1941 when 200 bombers dropped 650 HE bombs and 170 sets of incendiaries. Sporadic raids continued until early July 1941, after which they stopped until 27 July 1942 when Birmingham was raided by at least 60 planes that caused about 283 fires across the city. The last main raid on Birmingham took place three days later when incendiaries 'came down in showers' and 'there were grievous happenings in poorer quarters where houses hit by the explosive bombs collapsed on the inmates'. All told, the Blitz killed 2,241 Brummies, seriously injured 3,010 and harmed slightly 3,682. *The Blitz On Birmingham* by Carl Chinn.

24   31 Wellingtons and eleven Whitleys were dispatched on 21/22 October to Cologne, Hamburg, Reisholz and Stuttgart. The Wellingtons were sent to Hamburg, attempting to bomb the battleship Bismarck. Twelve fires were caused of which eight were classed as large in the Hamburg report. One Whitley was lost. *The Bomber Command War Diaries.*

25   *Wellington N2373 KX-K was abandoned near Blidworth in Nottinghamshire. Pilot Officer Milosalav E. Vejrazka the second pilot was killed. The rest of the crew were Pilot Officers L. Anderle, Ritcher and Furbach and Sergeant Landa.*

26   Two Hampdens - one on 49 Squadron and one on 83 Squadron – both at Waddington, FTR when they crashed in the North Sea. All nine crew men perished.

27   *Coventry Blitz* by H. C. Legg quoted in *70 True Stories of the Second World War* (Odhams Press Ltd).

28   In the last of a series of 'experience raids' designed to blood new aircrew, nine Wellingtons attacked the Porto Marghera oil refinery and storage depot in Venice. A secondary target was the oil-storage installation used by the Italian navy on a small island in the Venice lagoon. Three other Wimpys flew to Regensburg on the Danube, a port for Rumanian oil distribution. Affectionately known as the 'Wimpy' after the American Disney cartoon character J Wellington Wimpy in 'Popeye' who was always eating hamburgers and said, 'I'll gladly pay you Tuesday', the Wellington had been conceived by R. K. Pierson and his team using geodetic or lattice work structure developed by the brilliant British scientist, Dr. Barnes Wallis.

29   *Where The Red Ribbon Stopped* by Jon Rootes DFM quoted in *'Ordinary People'; True Accounts from wartime ex-aircrew* edited by Cyril Thompson (Barny Books 1999).

30   HQ Bomber Command sent messages of congratulations to the squadrons involved in the raid on Venice and HM King George VI congratulated crews when he visited RAF Mildenhall on 18 January to decorate a large number of airmen and officers including 'Speedy' Powell. The Royal party inspected the operations room and the photographic section and made a tour of the aerodrome, lunching in the Officers' Mess. When visiting the Airmen's Mess the King and Queen saw that the men were being entertained by a string orchestra as they ate! An enemy aircraft appeared overhead and was engaged by the station defences whilst the Royal party was still at the Station. No bombs were dropped.

31   Sergeant R A Hodgson's crew were all taken prisoner.

32   Quoted in Experiences of War: The British Airman by Roger Freeman (Arms & Armour 1989).

Chapter 3

German Defences

At nine o'clock on the morning of 16 April 1941, Lord Josiah Stamp, President of the London Midland and Scottish Railway (LMS) and a member of the Railway Executive Committee, arrived at his London office as usual, having stayed overnight at The Grove, the LMS war-time headquarters in Watford, He had just been appointed a member of the Transport War Council set up by the Minister of Transport. By the afternoon he was working in his office at Euston. In the early evening he travelled home to his wife and family at Shortlands in Bromley, Kent. That night the house took a direct hit from a bomb. Lord Stamp, his wife and his eldest son were killed, apparently drowned when their underground shelter was flooded, The deaths of Lord Stamp and his family symbolised the narrow gap between existence and oblivion, That he had supported appeasement before the war and had been an honoured guest of Adolf Hitler at the Nuremberg Rally proved no insurance against the sense of existential absurdity brought on by the randomness of war.'
Martin Bashforth of the National Railway Museum in York.

'As you already know' a sergeant pilot of Bomber Command told listeners to the BBC that first week of January 1941 'the RAF last week bombed military objectives in Bremen on three successive nights. I was on two of the three raids.' With the New Year, on the night of 1/2 January 1941, an attack of the same kind as the attack on Mannheim in December, but even heavier, was made against Bremen. Twenty thousand incendiaries, all dropped between an hour after nightfall and midnight, started enormous fires and made a mark for the high-explosive bombs. The objectives were substantial areas of the town, areas where there were shipbuilding yards, warehouses, oil refineries, factories - including the Focke-Wulf Airframe Factory - and railway yards. The weather was exactly right for the attack and there was snow on the ground besides, which made the rivers and waterways, black against the white earth, into most useful landmarks. The next night the attack was continued, but with rather fewer bombers; it was also less concentrated in time and the aircraft came in at intervals throughout the night. Yet a third attack was made the next night and this was all but as heavy as that of the first night. Here again the damage done, as was learned from photographic reconnaissance as well as from other sources, proved the effectiveness of concentrated attack.

'I will describe my second raid on Bremen' continued the sergeant pilot of Bomber Command 'because it is more vividly in my mind; in fact, I've never seen anything like it before. When we left there were so many fires in Bremen that I gave up counting them.

'We set off soon after tea-time: actually, I remember it was five-forty-six when we got on to the flarepath. The weather was very bad. We circled the aerodrome to get a bit of height: then almost immediately we had to climb through thick cloud, so that for the first half hour we were flying blind. Ice was forming on the windscreen and it looked as though we were going to have a bit of trouble so I started using my de-icers. But we came out of these bad conditions at seven thousand feet and above that it was beautiful weather - lovely and clear - with a quarter moon and plenty of stars. The only snag was that it was appallingly cold. In the ordinary way, I usually fly the aircraft out and fly it while we are doing the bombing: then when we've got clear of the target area, I hand over to the second pilot as far, say, as the North Sea: then I take over again. But this time it was so cold that we changed over several times both going out and coming back in order to try and get warm by moving about a bit. It'll give you some idea of what it was like, perhaps, when I tell you that once when I had some tea out of the thermos flask to try and warm myself up a little, there was ice at least an eighth of an inch thick round the rim of the cup when I'd finished drinking.

'But, apart from the cold, conditions were perfect. We got a bit of flak going over the Dutch coast but it wasn't really worth bothering about and I just carried on straight and level. So far as possible, I always try to keep a constant air speed going out and avoid jinking as much as possible - that's to say dodging about to avoid enemy fire - otherwise it only makes things more difficult for the navigator.

'Having got over this bit of flak quite safely we flew on steadily towards our target and without incident. When we were still about sixty miles away we could see a red glow in the sky and thirty or forty miles off we could actually see flames rising above Bremen.

'We went in about five miles south of the target. I cut the engines as much as I could to avoid being picked up by the enemy and of course, I was jinking now, all right.

'Then I did a very gentle left-hand circuit and I thought I'd try and count the fires as we were going round. I got up to thirty and then gave up. There must still have been at least another twenty or so in the target area.

'Some of the fires were very long in shape as though factories or other large long buildings were on fire. In one place there was a tremendous fire three or four hundred yards long. A deep red glow was reflected in the river and the fires were so numerous and so bright that they lit up the whole town and my navigator was able to make his run up guided by the bridges across the river. As we were circling I saw three sticks of bombs go down.

'Then, just as we were starting our run up, the flak became rather more heavy ahead of us and I saw another bomber running up, right across our track, but about a thousand feet above us and roughly half a mile ahead. This other plane was held in a large cone of searchlights: there must have been twenty-five or thirty lights on him. I could see his bombs leave the aircraft and I was able to follow them for about five hundred feet as they fell. I didn't see them burst, though, because I was too intent on our own run up, but out of the corner of my eye I could see this other machine trying to get out of the searchlights: just wriggling one way and another. It was an amazing sight. Finally, the pilot did a stall turn and dived out of the lights. It was very good to see him get out of them so neatly. Then we did our bombing, mostly with incendiaries. I did another circuit to see what results we'd got and saw additional fires starting and several heavy explosions where our bombs had fallen. Altogether we were there about twenty minutes, during which time I'd seen four people bomb beside myself: that's to say a stick of bombs every four minutes. By this time the AA people had got our height fairly accurately and we had one or two bursts very near which rocked the plane, but we got out of it all right and started for home. When we were about ten miles away I turned to give the crew a look back at Bremen. The whole place seemed ablaze. Then again when we were about a hundred miles away I turned again and even from that distance I could still pick out a red glow in the sky from the direction of Bremen. I've been on fifteen raids, including Düsseldorf and Mannheim and Berlin, but I must say I have never seen anything like this one before.'

Other major attacks of the winter and spring of 1941 were on Wilhelmshaven, Hanover, Kiel and Hamburg, while the offensive as a whole, against warships, naval bases and aerodromes, as well as against industries and communications, went on with steadily increasing strength and without a pause, except on those nights when the weather was hopeless.

Flight Sergeant George P. Dove, a Whitley air gunner on 10 Squadron at RAF Leeming recalls: 'Any trip over the other side was mostly tension, apprehension and then a sense of relief that another one was over. But our crew did one trip that left us chuckling for days afterwards. At the time crews were briefed for a primary and secondary target and if neither could be found any target of opportunity could be bombed. On this particular night we had been given two targets north of the Ruhr and duly took off at dusk on a fine evening, with a good weather forecast ahead. As they always did, the good people of Leeming village were dotted along the hedge bordering the airfield and waved as the laden Whitley lumbered along the grass, (no runways yet, as we had opened the brand new station just a few weeks previously). With the tail up and the nose almost touching the deck, which was characteristic of the Whitley (something to do with wing incidence I believe), we lifted over the boundary, did a circuit and headed out towards Hornsea and the long haul over the North Sea. Once well clear of the coast I gave the usual testing burst from the rear turret and then settled down till landfall over the Dutch coast. Apart from the usual welcome from the flak ships, we crossed the coast without incident and set course for the primary target. Despite the good met forecast we soon found ourselves in 10/10ths cloud (sounds familiar!) and any hope of finding either target was soon dashed. After sniffing around for an hour looking for a break in the soup, the skipper called up and said; 'That's it, we'll head North West over Holland and home.'

'The thick cloud persisted until we were well over Holland when it suddenly cleared and there, right ahead of us, was an airfield fully lit up, with aircraft circling with nav lights on. The navigator took a quick look and checked his map and shouted; 'It's Schiphol, the Luftwaffe is doing circuits and bumps.' So it was, we could scarcely believe our eyes. We waited for the challenge and the burst of light flak, but nothing happened. The skipper told us he was going down to take a closer look, warily we began losing height, expecting any moment to be on the wrong end of a burst of coloured tracer, but nothing! We were now down to 600 feet and we could quite clearly see six German aircraft on the circuit and one taking off. Everything was switched on - runway, perimeter and hangar lights - they must have been very over-confident. It was at this point that our pilot, who was a pre-war regular and very much a press-on type, decided to switch on our nav lights and join the circuit.

'Picture, if you will, a Whitley on circuit with assorted Luftwaffe aircraft over a German airfield. You could de-synchronise the engines of a Whitley and the out of phase Merlins did sound rather Teutonic and so I must assume this is why we were completely undetected. We must have arrived towards the end of the nights' exercise, because one by one the aircraft landed, until there was only one left on the circuit, us! The runway controller was flashing us a persistent green to come in, we were astonished at our good fortune. The skipper had by now decided on a course of action. He called on the intercom; 'I am going to do a long downwind leg and then come back over the hangars low and fast. Navigator, drop the bombs on the hangars and rear-gunner spray the airfield as we pass.' By chance our bomb load were all delayed action, perfect for the situation. We raced back over the airfield as fast as the Whitely could go, dropped the bombs and as the airfield came into view I pressed the triggers on the four-gun turret and kept them pressed till we reached the other side.

'As we sped out to sea, I gave a running commentary on what was happening, all the lights went out at once, bursts of flak, searchlights and red flares - all too ' late - we were well on our way home. When we landed back at Leeming and lit our first fag we were all elated and amused and told our story to a disbelieving de-briefing officer. Although we never heard any more about it, I do hope those bombs went off. It is easy to imagine a 'Mossie' getting away with it, but a Whitley?'

In 1941 the Air Ministry published an account[33] of a typical operation flown by Bomber Command, to Gelsenkirchen on the night of 9/10 January when 135 Wellington, Blenheim, Hampden and Whitley crews were dispatched. The action moves to a Wellington station - most probably Honington in Suffolk, home to 9 Squadron in 3 Group but as with every operation, it began at the very top at Bomber Command Headquarters at High Wycombe in Buckinghamshire: 'First, let us come with the Air Officer Commanding-in-Chief into his operations room at the Headquarters of Bomber Command. It is nine o'clock in the morning. Beneath a grassy mound, protected by deep layers of concrete, lies the place where he will order the forthcoming attack. Whether sunshine or rain prevail outside, inside downstairs there is always the soft light of a spring day shining steadily from half-concealed reflectors upon an oblong room of lofty and gracious proportions. It is air-conditioned, floored with rubber and entered by a single door only. This door and the stairway leading to it are guarded by sentries and no one lacking the proper authority may pass in or out.

On the main wall opposite the door there are three blackboards each about 30 feet by 10 feet. These display the Order of Battle. The Commander-in-Chief [Air Chief Marshal Sir Richard E. C. Peirse KCB DSO AFC who as Vice-Chief of Air Staff had already been closely concerned with the British bombing policy and had been appointed C-in-C Bomber Command on 5 October 1940][34] has only to glance at them to see at once the exact strength of every Group, the whereabouts of the squadrons in it and the total number of aircraft available. The left-hand board is devoted to current operations. It shows what Groups are carrying out what tasks and what targets were chosen for attack on the previous night. Information upon it is written in chalk in two colours; that inscribed in yellow shows what it was decided to do, that in red what was actually carried out. The information on the other boards is displayed in material more durable than chalk and is kept up to date once every twenty-four hours. Above these boards, which occupy the whole of one wall, is a clock and below it the date displayed in large letters and figures. The right-hand wall of the oblong room is covered by a meteorological map showing the state of the weather. The data displayed on it are changed every eight hours. Beside it is a moon chart recording the periods of moonlight and darkness throughout the month. Opposite on the left-hand wall is a quarter-inch map of Northern Europe, showing the main targets. Their positions are marked by pins with coloured labels attached to them on which the code formula for each target is written. On the back wall is a similar map displaying the main targets in Italy.

To the left of the door, a short distance out from the wall, are the desks of the controller and the duty officers. Telephones on these desks connect Headquarters with the Groups and the Air Ministry. In the corner to the right is the desk occupied by the naval staff officer attached to Bomber Command. It is his duty to advise the Commander-in-Chief on all naval matters and to keep before him the views of the Admiralty. He is a captain in the Royal Navy and has two assistants, a lieutenant-commander, who is usually stationed with the Group operating against enemy shipping and another working mostly with the Group engaged on mining. There is also an Army officer who maintains liaison with the Commander-in-Chief of the Home Forces.

In the left centre of the room stands the desk of the Commander-in-Chief. Near it are three large tables mounted on pivots so that they can be moved at will from the horizontal to the perpendicular. On the first of these are pinned the maps in current use for the night's operations. There is also a photographic mosaic of the whole territory of the Ruhr. On the second table there is a large map of Europe showing the routes to the various targets and places where German night fighters seek to intercept our bombers or where they are known to patrol. The data on this map are changed every twenty-four hours in accordance with information supplied by the Air Ministry. On this table, too, are a number of target maps grouped round a large-scale map showing where the targets they depict are situated. The third table displays graphs from which can be learnt immediately the number of times various classes of targets have been attacked. There is also a map of Berlin and enlarged photographic reproductions of the more important targets.

Seated at his desk the Commander-in-Chief makes a rapid appreciation of the situation from the reports of the previous night's operations and from information which may be supplied to him verbally by the senior air staff officer and the group captain in charge of operations. The type of target to be attacked is already known. It has been chosen by the War Cabinet which determines the major direction of our air offensive. The Air Ministry issues directives from time to time based on the decisions of the Cabinet. It is the duty and responsibility of the AOC-in-C to implement these general directives.

Before he can choose the targets for the night he must consult meteorological experts, for his choice is dependent on weather and visibility. These experts are civilian members of the Air Ministry and serve in all the Groups and Stations throughout the Command. Their head awaits him beside the board on which the picture of the weather is displayed. It is not always possible for the met officer to give a final forecast of the night's weather at nine o'clock in the morning. He is, however, in a position to indicate the general trend over more than one area. The final forecast is made sometimes as late as 4 o'clock in the afternoon, after the mid-day telephone conference has taken place between the Group met officers.

It is not unusual for the Commander-in-Chief to choose alternative areas and to issue orders for plans to be made to attack targets in all of them, for, when the weather is uncertain, he may reserve his final decision until he has obtained the final forecast. Thus, at his 9 o'clock conference he chooses a primary and a secondary target, the secondary only to be attacked if the weather over the primary should prove to be unfavourable. Only places that can be attacked by the same type of bombs as those selected for primary targets will be chosen as secondary targets, since a change-over in the type of bombs to be carried may interfere considerably with the 'bombing-up' of the aircraft and lead to delays.

The met officer delivers a short lecture on the weather, which may last some ten minutes. It is his practice to be as definite as the vagaries of his subject allow. He talks not only of the weather on the route, but of that which our aircraft will find over their aerodromes on their return.

The Commander-in-Chief returns to his desk with his mind made up concerning the areas containing the possible targets for the night's attack. Before he chooses the actual targets he may call for information from the chief intelligence officer and the group captain in charge of operations. Photographs of different targets are brought to him. Then he makes his choice.

Having decided the target, the number of aircraft to carry out the attack is next discussed. The controller has been getting into touch with the Groups and within a few minutes the exact numbers of aircraft available in each Group are placed before the C-in-C who then decides on the proportion of heavy and medium loads to be carried. Finally he discusses with the senior air staff officer the number of aircraft which can be put over any one target, during the number of hours of darkness available. This is a matter of great importance and one which those who cry out for an overwhelming attack on a single objective are apt to forget. It is never possible to concentrate effectively more than a certain number of aircraft over any one target if the night is short.

This point decided, the C-in-C takes up a form marked 'C-in-C's Daily Allotment of Targets.' On this he writes down the code formula for the targets to be attacked, the number of aircraft in each Group to take part in the operations and the proportion of incendiary and high explosive bombs to be carried. This sheet constitutes the written order for the operation. It is passed to the Controller, who at once issues the necessary orders to the Groups. The Commander-in-Chief then returns to his office. The process has taken less than an hour and in that period it is possible for him, if he so desires it, to plan the despatch to any target of any fraction or of the whole of the bomber strength available.

The scene must now be shifted to a Group Headquarters Here, on the outskirts of some town in East Anglia [3 Group Headquarters was at Exning House near Newmarket], the Midlands or the North of England the Air Officer Commanding the Group receives the orders. These are sent to him on the direct telephone line and by teleprinter. He knows the number of aircraft he is required to despatch on the night's mission and he then decides how many squadrons he will use and what stations he will operate. Much the same procedure as that which has been followed at Headquarters, Bomber Command, is now carried out at Group Headquarters. They despatch the orders and indicate the targets to the commanders of the Stations chosen for the night's operation. The Station and Group Met officers next meet in conference over the telephone.

At Station Headquarters, the scene is still the same, but the scale is smaller. The Station commander, a group captain, sends for his squadron commander and operations officer. He repeats the orders he has received. The aiming points on the target maps are marked in readiness for the night's operation.

The squadron commander faces one main problem: How can the allotted target be most easily located and identified? This depends primarily on the conditions of visibility. Is there enough moon? In that case it should be easy. Is it a dark night with clouds? In that case the aircraft may have to spend upwards of an hour in the target area before the crew can make certain that they have found their mark. Then there is the distance between the base and the target to be considered. Sufficient petrol has to be carried for the journey out and back and a margin provided for the time spent over the target and for the possibility that cloud or fog over the base aerodrome may make it necessary for the aircraft to be diverted to more distant aerodromes. A vital factor affecting the amount of petrol to be carried is the course to be taken. This is often not direct, since the attacking aircraft must avoid, if possible, areas where flak or searchlights have been concentrated. There is, for example, a searchlight corridor in Northern Germany that is well known to the Royal Air Force and has to be reckoned with every time targets at Bremen, Hamburg or in the Ruhr area are attacked.

The size of the bomb loads is laid down in the Group Operation Orders, but the load may be reduced by the Station commander if he thinks it necessary to do so for local reasons. It must be emphasised that, once the target has been chosen and the aircraft 'bombed up,' to change it at short notice, although not impossible, is difficult. It means changes in the fuel and bomb load. These cannot be made in a few minutes and if the decision is left too late, it may mean that an unsuitable bomb load will be delivered at the target.

Let us now take a glance at the Station itself. It is in most cases of recent construction and the layout follows up-to-date principles. The buildings, camouflaged so as to cause them to blend as far as possible with the colours and contours of the surrounding country, are constructed in blocks with considerable space between them and cover a wide area. A network of roads connects them with Station Headquarters where the Operations Room is situated, with the Officers' and Sergeants' Messes, with the quarters of the men, with those of the WAAF, with the hospital and decontamination centre, with the bomb dump, with the hangars and with the airfield itself, which is surrounded by a perimeter track. For obvious reasons aircraft no longer live in hangars. They are dispersed round the field in such a way as to minimise any effects which may be caused by bombing. They remain and are serviced in the open air, only being taken to the hangars for some major repair.

It is the duty of the ground crews to keep the aircraft serviceable. Their importance is, therefore, very great. The success of the attack and the lives of the flying crew depend in the last resort on their labours. As soon as an aircraft has landed from a sortie the night staff on duty cover its engines and turrets. At daylight the ground crew, consisting of two fitters and two riggers, go over the whole aircraft from nose to tail. The oil and fuel consumption of the engines is checked against the pilot's log. A special watch is kept for oiled-up sparking plugs. The wings and fuselage are checked for holes caused by anti-aircraft or machine-gun fire. The bomb racks are examined and an electrician checks all the electrical gear, paying special attention to the bomb release gear. The controls are tested. The tyres pumped up. The whole process takes from two to three hours.

In addition to the ground crews there is the personnel of the WAAF. The number of these stationed at Bomber Stations is steadily increasing. The Women's Auxiliary Air Force carry out many of the routine duties - cyphering, orderly room work, teleprinting and telephone operating, driving, cleaning sparking plugs and cooking. Their work is varied and performed with great efficiency. They are rendering an essential service, often in places of great danger. They have their own quarters and are under their own officers.

As soon as the preliminary orders for the raid have been received, the work of fuelling and 'bombing-up' is put in hand. The flight-sergeant in charge of the bomb dump and his staff make up the different bomb loads and the yellow-painted bombs are loaded on to trolley trains drawn by tractors. These trains visit each aircraft and the bombs are transferred from them to the bomb hatches. Portable cranes are used and the bombs, which have already been fused at the bomb dump, are slung into position. They are attached to the aircraft by means of lugs and released by an electro-magnetic system which is controlled by switches operated by the bomb-aimer. The fitting of the bomb accurately into its rack calls for time and trouble. The lugs attached to the bomb must be correctly aligned with the lugs in the bomb hatches; otherwise the bomb will not fit. To do this in a hurry with bombs weighing a thousand pounds or more is not easy and requires skill, practice and team work. An expert 'bombing-up' squad of twenty-eight men can load fifteen aircraft in two hours.

Now come back to Station Headquarters. The crews, who have had a preliminary warning that they will be wanted that night, are assembled in the Briefing Room some hours before the start of the raid. They sit facing a dais behind which is a blackboard. In many such rooms the following notice will be found on one of the walls: 'It is better to keep your mouth shut and let people think you're a fool than to open it and remove all doubt.'

Once seated, the crews are told what they have to do that night. Here is an actual briefing for an attack on an oil target. It is the squadron commander or the intelligence officer who is speaking: - 'The target to-night is the synthetic-oil plant at Gelsenkirchen. There are two main types of oil plants in Germany: oil refineries for treating crude oil, imported or home-produced and synthetic-oil plants. The Gelsenberg-Benzin A-G, which is our target, consists of two atmospheric distillation units producing petrol from coal.

'The development and extension of these works was undertaken in 1938. Their output capacity is 325,000 metric tons per annum. The most vital section of this plant and also the most vulnerable is the hydrogenation plant itself-marked 'B' on the illustration. It lies in the top half of the target running from the narrow-neck in a north-westerly direction and covers most of that part of the target. This section of the plant consists of the following principal components:-

(1) Compressor house.
(2) Hydrogenation stalls.
(3) Water-gas units.
(4) CO conversion plant.
(5) Sulphur purification.

'A direct hit with a large bomb on the compressor house, where the low pressure hydrogen lines lead to the compressors, will cause a real explosion which is always likely to lead to severe damage in a building containing a large amount of moving machinery.

'Damage to the compressors will put the whole plant out of action and, since they are most difficult to replace, the compressor house would be a very profitable aiming point. A hit here may cause damage out of all proportion to the size of the bomb.

'The plant lies on the northern bank of the Emscher Kanal, which at this point runs parallel and very close to the Rhein-Herne Kanal.

'The main town of Gelsenkirchen is on the south bank, but to the west of the target is an industrial residential area.'

[A description of the landmarks by which the target can be found is then given and the suggested lines of approach indicated.]

'Get right up to your target and do your stuff.

'Note carefully and report the position of any outstanding landmarks for future reference, also the position of flak and searchlights. There is a strong concentration of lights reported just north of the town.

'The cameras are on aircraft, letters A, B, C, D. Get us some good pictures.

The leaflets are on aircraft, letters Z, Y, X, W, V, U and are to be dropped in the target area.'

'The approach to the target is suggested but not laid down in a hard and fast manner. Captains of bombing aircraft are allowed considerable latitude in the choice of routes to the target, once the area in which it is located has been entered. This is natural, for it is impossible to foresee the exact circumstances in which they will be called upon to make the attack.

Particulars are then given, based on the 'opposition map,' of what enemy defences may probably be encountered. These are of three kinds: night fighters, anti-aircraft guns (flak) and balloon barrages. The crews are shown on a map the area in which the night fighters operate, the places where they have already been encountered and the position of balloon barrages.

The navigators, who are also the bomb-aimers, are then issued with target maps. These are all that they take with them to help them identify their targets. The maps are simplified to the greatest possible degree and are printed in various colours which represent respectively woods, built-up areas, water and other easily distinguishable features. The target itself is clearly marked in red or orange. Photographs of the target are also shown to the crews and enlargements may be thrown on a screen by means of a projector.

The crews are then addressed by the signals officer who informs the wireless operators of the frequencies to be used for normal traffic, for identification and for 'fixes' (signals sent out by an aircraft to enable its position to be determined by the special services established in this country for the purpose). The homing and distress procedure is also explained. The procedure connected with use of wireless is constantly changed. The tendency at the moment is for fewer and fewer calls to be made on wireless, which is coming to be used only in moments of real emergency. Navigators are expected to find their way to and from the target by other means and the number of those who do so is increasing daily. It has come to be a point of honour with them not to ask for directional bearings, for it must always be remembered that the ideal bomber crew is the one who succeeds in going to the target and back again in the shortest possible time without being detected by the foe. Warnings, sometimes couched in solemn terms, are given against making use of the wrong procedure when in distress. Here again much depends on the training of the crew. The position of our own searchlights is explained, together with the method of obtaining their aid in case of emergency when the aircraft is nearing home.

Then the met officer takes up the tale. After he has finished the armaments officer delivers a short disquisition on the nature of the bombs carried and their fusing.

The briefing may last as long as three-quarters of an hour; but usually it is shorter. The atmosphere in which it is conducted resembles nothing so much as a lecture at a university, though the attention paid by the audience would certainly flatter most lecturing dons. Everything is very matter-of-fact. There is no straining after effect. The information is imparted clearly, briefly and without embellishment. Questions are answered in the same way. The object; aimed at and achieved, is to leave no member of a crew with the excuse that he did not know that a certain procedure was to be employed or a certain course to be avoided.

After the briefing, the captains of the aircraft detailed will spend some time with their navigators working out the best course to and from the targets within' the limits set. The captains will then check over the aircraft with the ground crews to make sure that everything is working. The aircraft has been previously taken off and flown round the aerodrome for about half an hour. The instruments which have to be most carefully watched are the electrical and hydraulic. A defect, for example, in the inter-communication system makes it impossible for the captain to communicate his orders to the crew and, though in other respects the aircraft may be perfectly serviceable, it is useless for operational purposes.

The navigator takes on board with him a green canvas satchel in which he keeps all his gear. It holds signal cartridges for the Very light pistol, message pads, a dimmed torch and ' flimsies' on which is typed the procedure to be adopted if the aircraft is lost and requires wireless assistance. The 'flimsies' are made of rice paper, so that in the event of emergency they can be destroyed by being eaten. It is said that the taste of the ink leaves much to be desired. In another pocket a protractor, dividers, coloured pencils and a course and speed calculator are stowed; in another, a log book, target map and a questionnaire to be filled in if the navigator in addition to dropping bombs is also going to take photographs. There is also another shorter questionnaire to be answered for the benefit of the met officer, who is thus enabled to check what the actual weather was like during the flight and over the target and compare it with his forecast. Finally there are the Astro navigation tables. Astro navigation has been brought to a high pitch of proficiency, though it can only be used in certain types of aircraft and, of course, only when the stars are visible. Five to eight miles is the maximum margin of error if the navigator is skilful.

The crews then have a meal, after which they put on their flying clothes. These are of great variety and are worn over their uniform. The issue of flying uniform cut to much the same pattern as the Army battledress is becoming more general. Over this a sweater may be worn and then the Irvine jacket, which contains the 'Mae West.' This can be inflated instantly on reaching the water. On the feet silk socks are worn and over them woollen stockings and flying boots lined with lamb's wool. The flying helmet contains the oxygen mask with a tube that can be plugged into the oxygen supply. It is not usual for the crew to wear their parachutes, but to keep them handy clipped to hooks near their stations. The pilot, however, usually sits on his. Just before leaving, the crew have each been issued with a paper bag containing the rations for the flight. These vary according to what is obtainable: the ideal ration consists of a few biscuits, an apple or an orange, a bar of chocolate, some barley sugar, chewing' gum and raisins. Each member of the crew also carries a thermos flask of hot tea or coffee. The crews are conveyed to their aircraft, dispersed round the aerodrome, in a lorry. On reaching it they get aboard the aircraft. Let us go with them.

Though the bomber has looked huge enough on the aerodrome in its coat of dull black paint enlivened only by the code letters painted on it and by some fanciful device chosen by the squadron or by the individual captain - a bent bow with the arrow against the string, a large portrait of 'Jane' of picture paper fame, a bird with spread wings; a kangaroo on a cloud - inside there is very little space. The air-gunner crawls into his turret and closes the door. The wireless operator goes to his' cabin and the navigator to his, the pilot and the second pilot to their controls. If you are to walk up the fuselage, you must bend your head. There are guy ropes to hold on to.

Once at stations, the first task is for the wireless operator to check his wireless. He does this by speaking to the Watch Office. All wireless communications from the operating aircraft are received at this office, which corresponds to 'the control tower at a civil aerodrome. In it there is a radio receiving and transmitting set and from that' office the' departure' and arrival of the aircraft are controlled. Each aircraft, in addition to the code letters showing the station to which it belongs, has its own individual letter and is known by that letter throughout the operation. The station itself has a code name. The usual way of checking the wireless is, therefore, as follows:- 'Hullo Parsnip (code name of the station), Hullo Parsnip, 'E for Edward' calling, 'E for Edward' calling, are you receiving me, are you receiving me? Over to you, over.'

If all is well the Watch Office will reply: 'Parsnip answering 'E for Edward', Parsnip answering 'E for Edward', I am receiving you loud and clear, I am receiving you loud and clear, strength niner (nine).' It will be observed that all signals are repeated sentence by sentence to be sure that they should be properly understood.

The aircraft are sent off at short intervals of between two and five minutes. The signal to take off is made to them by those in charge of the flare path who flash a green or red light indicating whether the aircraft should or should not begin its run. During the periods of take-off and landing, an ambulance and a fire tender stand beside the Watch Office ready for emergencies. On receiving the signal to take off, the pilot opens up his engines. He may keep his brakes on to lift the tail before starting the run. The Whitleys when fully loaded weigh about sixteen tons and require about 1,000 yards run to take off. The Stirlings, Halifaxes and Manchesters weigh much more, but take about the same amount of run. The pilot has his wing flaps slightly lowered. As soon as the aircraft is airborne, the wheels and flaps come up. On reaching 1,000 feet course is set for the objective. If the wind is favourable; the aircraft flies straight from the take-off on to its course: If not, it will circle the aerodrome. The captain has to make quite sure that his aircraft is setting off dead on the right course from the start; consequently the first words spoken on the intercom are usually by the navigator. He will say: 'Hullo pilot, the course is X'.' As soon as the pilot has turned the aircraft on to that course, he reports: 'Hullo navigator, on course.' After that there is generally silence except for the navigator. He may ask the pilot questions in order to check the height and speed of the aircraft during its run to the coast.

On reaching the coast, the navigator pin-points' his position and, if the aircraft is slightly off course, he gives the necessary directions to bring it dead on course again. As soon as the coast is left behind, the pilot begins to gain height and will say to the crew: 'I am going up now to X feet, speed so and so.' Once the course is set, the navigator is left in peace as much as possible to carry on his difficult task. As has already been said, he will work out the course by means of the stars and also by a form of wireless directional aid which can be used while keeping wireless silence. Over the sea the bombs are made 'live.' Inside the aircraft there is darkness. If the crew wish to see, they use hand torches suitably dimmed. The wireless operator has an amber light to enable him to make his entries in the log which he must keep. The captain can often be heard giving the order: 'Keep your lights down.'

Now come back along the fuselage to see what the rear gunner is doing. He has settled down in his seat; his parachute is hung up behind him; he has locked the turret doors. The turret is power-operated and swings easily in any direction. First he tests it, moving it to and fro by pressing on a pair of handles rather like bicycle handles. He loads and cocks the guns. This done he switches over his intercom and reports to the captain that everything is working.'

The rear-gunner on a 9 Squadron Wellington at Honington, who was introduced to listeners as 'a commissioned air gunner in the RAFVR, 'a flight lieutenant aged 39 and a well-known big-game shot' - spoke to the nation on Winged Words a BBC broadcast about the raid on Gelsenkirchen on the night of 9/10 January.

'In our crew the captain and second pilot were Scots; the two wireless operator air gunners were from Canada and the Irish Free State, while the navigator came from the West Indies; and I'm an Englishman. One of the gunners is young enough, with due precocity on my part, to be my son.

'A tail gunner is part of a crew and this crew's life dominates not only his flying hours but his whole existence. Crews are married up at an operational training unit, or on arrival at their squadrons and after that they are never parted. Crew life becomes unendingly intimate. On the trip, you don't see them at work, but you know they're there and you take comfort from each other. Without being sentimental, there is a sense of comradeship about the venture. You come together, six nondescript individuals - young and old, lean and fat, officer and non-commissioned officer. You eye each other in a rather British sort of way and wish you could find something graceful and appropriate to say. You can't. You think how odd they look and I suppose you must look just as odd to them. None of you would probably have chosen each other if crews were made on the picking up principle, but after a bit you would not dream of changing. It is really very curious.

'The striking thing about a tail turret is the sense of detachment it gives you' - it is a speaking - 'You're out beyond the tail of the plane and you can see nothing at all of the aircraft unless you turn sideways. It has all the effects of being suspended in space. It sounds, perhaps, a little terrifying, but actually it is fascinating. The effect it has on me is to make me feel that I am in a different machine from the others. I hear their voices. I know that they are there at the other end of the aircraft, but I feel remote and alone. Running my own little show, I like to sense that they are able to run theirs feeling that they needn't worry about attack from the rear. Some gunners have told me that this sense of isolation weighed heavily on them at first, but I have spent a lot of time occupied with solitary pursuits and it has never irked me, personally ...We must keep a good look-out, you and I, in our rear turret to-night, for, in the last month or so, the enemy fighters have been more active by night; and quite a few of our gunners have been engaged. Previous to that we had, unfortunately, not had much opportunity of using our guns, except during the period of the fighting in France, when we got quite a lot of good ground targets at low altitudes. I remember with peculiar satisfaction a long white road in Northern France, a full moon and a German lorry column, a particularly desirable combination, if I may say so. But from a gunnery point of view our outings have often been, as Dr. Johnson said of second marriage, 'the triumph of hope over experience.' Now we are rising slowly over the familiar, darkened landmarks below. A pause and we have crossed the coast and we ask the captain's permission to fire a burst into the sea, just to make assurance doubly sure as regards the serviceability of our guns. Out at sea, away on my beam, I suddenly see another aircraft, a twin-engined plane flying parallel to us. It is a long way off. Can it be a Messerschmitt 110? I report to the captain and keep it in view, but as it swings in I recognise the high familiar tail fin of the Wellington. Soon it has disappeared again in the darkness. 'Good hunting'.'

Generally speaking, crews do not talk very much over the intercom. They are too much occupied. Besides, they wish to conserve oxygen as much as possible. The oxygen is turned on when the aircraft has crossed the enemy's coast.

Time passes. The aircraft is now over the Dutch coast, perhaps above a bank of clouds. If these are thick, the navigator will find his course by the stars or by the burst of flak; if the night is clear, by the water landmarks of Holland which cannot be disguised. 'Soon we are flying high above a bank of clouds' continued the rear gunner. 'It is lit from below by German searchlights and this gives it a sort of opaque glow. Our captain comes down just above it, so that we can have cover if it is needed. Ten minutes later we are past the clouds, climbing again. We have been this way before and we are getting to know it quite well. Now the Germans are after us with their searchlights - and pretty good they are, too. Out in front there is a flak barrage, otherwise known as an anti-aircraft barrage. In the tail turret you can't see the barrage yet. The searchlights keep crossing and crossing. Now one's caught us. But no. After holding us for a moment it passes. Two minutes later, however, they get us good and proper. And very confusing it is, too. You feel a cross between a fly on an arc lamp and a man whose clothes have been pinched while he was bathing. But, of course, it's a good deal worse for the captain, who's flying the aircraft.

'We turn and twist, hoping to get clear and now the party's starting! Here comes the flak. Personally the German flak has never worried me very much. Perhaps I've been lucky. You can see the pyrotechnics coming bursting up at you and going off all round you, with a sense of detachment. It's a Brocks' benefit - and all for you. It would cost you a shilling at the Crystal Palace. I have never really honestly felt it could be going to hit me. I suppose I'm the usual indolent English optimist. And if it does catch us, we have the benefit of our marvellously constructed machine. They stand a lot of punishment. A large hole was once made only four feet behind my seat and I never even knew the old kite had been hit.'

By this time the crew are keyed up, waiting for the moment when the bombs will be dropped. If the target is not easy to find, the pilot will be heard asking the navigator, who has now gone forward and is lying or sitting in the bomb-aiming position, his attitude depending on the type of aircraft, 'Is it time to turn in?' or 'Is it time to make our run?' Presently the navigator will say, 'OK, turn in.' By this time the captain will have decided whether to attack direct or whether to make a gliding attack. In any case he is probably taking avoiding action - 'jinking,' as it is called - and is flying with engines de-synchronised.

A gliding attack tends to confuse the defences, whose sound locators cannot pick up the aircraft. When making it the pilot announces his height every 200 feet. When about to turn in for the run the pilot will say 'Opening bomb doors.' This is done hydraulically. As soon as they are open the navigator takes charge. He brings the aircraft on to the target by instructing the pilot how to steer. If he wishes him to turn to the left he will say 'Left, left,' repeating the word. If, however, he wishes him to go to the right he will say 'Right' once only. The reason for this is that there is often a considerable amount of crackling on the 'intercom' which makes it difficult to distinguish the exact words spoken. If he hears two words, the pilot knows that they must be Left, left; if only one word that it must be 'Right' and he alters course accordingly.

A gliding attack lasts from four to five minutes. Presently the navigator will say 'Steady' and the pilot will then hold the aircraft on its course until he hears the navigator say 'Bombs gone.' All this time the navigator has been gazing through the bomb-sight. Conveniently at hand are a number of switches by which he can control the manner in which the bombs are dropped. These may fall in a 'close' or an 'open' stick; that is, he can either let the bombs go simultaneously, in which case they will fall in a bunch, or one by one at short intervals. The bomb-sight is so constructed that it can be automatically set to make allowance for the ground speed of the aircraft and the force and direction of the wind. If necessary the bombs can be released automatically when the aircraft has reached a certain position indicated on the bomb-sight. They are also released by hand by means of pressure on a button. As soon as they have fallen the navigator reports 'Bombs gone.' In a gliding attack the pilot will continue to glide in order to leave the target unheard, if possible. Before he can close the bomb doors he has to open the throttle.

'Well, we are getting pretty close to the target now' continues the rear gunner 'and I can hear the navigator and the captain chattering away over the intercom; but actually there is no need to worry about spotting our target to-night, because some more of our bombers have been there first and the factory we're after is blazing away nicely. It's a terrible temptation to the gunner to sit and watch the bombs dropping. But really he oughtn't to, because we may be attacked at any moment and the rear gunner's job is to watch for their attack, not ours. Still, let's have a peep or two out of the corner of our eye. The first stick seems a bit wide, but the second hits the target square as far as one can judge and adds to the blaze. 'Whoopee!' shouts the second pilot. 'Whoopee!' shouts back the captain: and 'Whoopee' I shout from the back.

'We waste no time but turn for home. This is where we may expect attack.

We have been fired at pretty continuously all the time we have been over the target area, but now the flak has stopped and there are only the searchlights. This seems to suggest fighters. A few nights earlier, in this same area, a machine from our squadron met an enemy fighter under just these conditions. With both aircraft illuminated by German searchlights, the fighter came bursting up and started banging off tracer at about 600 yards. It went low. Our gunner let him come to within three hundred yards and then gave him three or four bursts. He banked sharply and then broke away. However, the gunner thought that wasn't the end of him, nor was it. He came in again, slightly above and firing off red and green tracer with all the enthusiasm associated with the fifth of November at a prep school. This time our gunner gave him all he'd got. But he didn't need the lot. He just went into a vertical dive and pitch-forked himself into the Reich.

'Well, we're all teed up for something to happen; but it doesn't. More searchlights, more flak, but no fighters and in due course we are crossing the coast again, though that in itself spells no immunity from attack. It's beginning to feel pretty chilly because we have been flying at a good height; and I suddenly find that one of my legs is getting cramped and that six and a half hours of scanning the heavens has been a bit of a strain on the eyes; and that my hands have grown weary of holding the grips that operate the turret. In short, quite suddenly, one finds that a lot of time has passed, much to one's surprise and that one's feeling tired. Still, anything may happen at any moment, one keeps telling one's self - one must not relax.

'Now we're over our own coast. Searchlights catch us at once. Our searchlights are really good. We've had a good trip. Things have gone well. The target was found easily and was well and truly hit. There's a happy atmosphere inside the kite - though nothing is said. You notice the barometer rising. It's sort of psychological.'

Next door to the Operations Room at a Station is the Wireless Room, where a constant watch is kept. A wireless operator is always listening in on the frequencies allotted to the Station. Once near the home aerodrome, the wireless telephone comes into play and the aircraft is brought to land by the voice, from the Watch Office. A bomber pilot is trained in blind flying and in the art of landing along the Lorenz beam. A simple system of sound signals enables him to know exactly where he is, whether he can see the ground or not. The signals change as he gets nearer the aerodrome. Pilots carrying out this training, which is done with great regularity daily, are appropriately described as on the BAT flight, these letters standing for Blind Approach Training. Often in cloudy weather the pilot will ask for the barometrical pressure above the base to enable him to set his barometer and thus to know with very great accuracy what his height is above the surface of the earth.

Meanwhile, the staff of the Operations Rooms at the Stations, at the Groups and at Headquarters, Bomber Command have been waiting through the night. They plot results as they come in. These are recorded on boards in different coloured chalks or on charts by means of labels. Sometimes fog may descend on a base while the aircraft from it are still out on their sortie. Signals must then be sent, diverting them to another base where the weather is clear. Should an aircraft be in distress, the life-saving service is warned and if it comes down in the sea, the launches of the RAF and, if necessary, the local lifeboat goes at once to the rescue. The crews are provided with a rubber dinghy inflated automatically which can be shot out of the aircraft. They wear a yellow covering to their flying helmets which in daytime makes it easy to spot them from a height.

As soon as the aircraft arrives back over the aerodrome, the Watch Office sends a signal telling the captain when to land. If, as often happens, several arrive together, they are told the order in which to land and those waiting their turn circle the aerodrome at a certain height. The angle of glide indicator is set to one side of the flare path, the lights of which are switched on for each landing. They include a brightly illuminated letter 'T' which is situated at the beginning of the flare path. There is also the Chance light. It is so called after the name of its inventor, not because it is turned on haphazard. It illuminates the ground at the entrance to the flare path.

'Well, here we are, circling the aerodrome, waiting for permission to land' concluded the rear gunner. 'In we come - a good landing - and we taxi up to the hangar. The CO's on the tarmac - 'Square' by nickname and square by nature [Wing Commander Andrew 'Square' McKee, the thick-set New Zealander CO of 9 Squadron, who was about five foot five in both directions] - and he wants to hear about it; and then we go and pull off our flying-kit; swap a few experiences in the crew room; put in the report.' [35]

When the crews have landed they are taken in lorries from the dispersal point to the Briefing Room or to the Operations Room - the procedure varies with stations - where they are interrogated by an intelligence officer who has been present at the briefing and taken part in it. To ensure a certain uniformity in the reports, intelligence officers make use of a questionnaire. The crews are interrogated one by one as they arrive and the interrogation is thorough, even when they report ten-tenths cloud which has made it impossible for them to see the, target. Two or three intelligence officers may be employed on this if there have been a large number of crews out on the operation. The room is soon full of pilots, navigators, wireless operators, air gunners, sipping tea or coffee and answering questions. 'I dropped the incendiaries immediately north of the dry dock ...fires looked like a heath-fire to me... I couldn't see the explosions because of the searchlights... Over Berlin there were so many searchlights that they lit up the ack-ack bursts... 'two rows of very vivid bright fires...' the *Scharnhorst* doesn't look so *Gneisenau*... we did one of those horribly steady run-ups and I saw the bomb flashes... there was a wee spot of light flak here... when I dropped the new bomb from 12,000 feet the whole aircraft shivered as though a shell had burst near it... absolutely screeching round the sky... the moon was wizard.'

'And so to bacon and eggs and bed in the pale light of a dawn I used to associate with roe stalking and cub hunting, though that seems a long time ago now.

Only 56 aircraft reported bombing the designated targets and the Germans reported bombs dropping not only in Gelsenkirchen but also in adjacent towns of Buer, Horst and Hessler. One Whitley failed to return.

Endnotes Chapter 3

33   *NIGHT RAID. Briefing the crew - Bomber Command; The Air Ministry Account of Bomber Command's Offensive Against the Axis September 1939-July, 1941.*

34   Peirse was the son of an admiral and had fought with conspicuous gallantry in the First World War. He had held several colonial commands during the inter-war years before heading up the intelligence section of the RAF and becoming Deputy Chief of the Air Staff in 1938. The choice of German transportation as the main objective of the bomber force and the German morale as the secondary was a confession of failure. In January 1941 Peirse was instructed that the whole primary aim of the bombing offensive should be the destruction of the German synthetic oil plants. The only diversions contemplated from this strict programme were such operations as might be necessary against invasion ports and enemy naval forces. At the outset the oil campaign was curtailed by the weather and then it was discarded, temporarily at least when in March Prime Minister Winston Churchill had given absolute priority to the Battle of the Atlantic. By July 1941 Peirse had a nominal strength of 49 squadrons or almost 1,000 aircraft on paper but eight of these squadrons were equipped with Blenheim light bombers. Although there were eight squadrons of the new 'heavies', four were not yet operational. Only 37 of the 49 squadrons were thus available for the assault against German transportation and morale and even these could not play a full part as less than two-thirds of their crews were fully trained and ready for operations. This had arisen mainly because of the drain of experienced crews to the Middle East, together with a shortening of the courses in the OTUs in an attempt to gain a greater output of aircrews. So for the time being at least, Peirse had to rely heavily on the Wellington squadrons, for so long the back bone of Bomber Command operations. On 22 February 1942, having been recalled from the USA where he was head of the RAF Delegation, Air Chief Marshal Arthur Travers Harris CB OBE arrived at High Wycombe, Buckinghamshire, to take over as commander-in-Chief of RAF Bomber Command from Air Vice Marshal 'Jackie' Baldwin DSO OBE who since 8 January been standing in for Sir Richard Peirse, who had become AOC-in-C India on 6 March 1942. *The Right of the Line* by John Terraine (Hodder & Stoughton 1985).

35   Andrew 'Square' McKee was born on 10 January 1902 at Oxford, New Zealand. He joined the RAF in 1927 and served in pilot roles in India and the UK before joining the Air Staff at Headquarters 3 Group in 1938. Promoted Group Captain, McKee became the Marham Station Commander in 1941 before becoming AOC 205 Group in April 1945. After the war he was appointed Senior Air Staff Officer at Headquarters Middle East Air Force and was then made Commandant of the Officer Advanced Training School in 1947. He went on to be Commandant of the RAF Flying College in 1949, Air Officer Commanding 21 Group in 1951 and SASO at HQ RAF Bomber Command in 1953. His final post was as Air Officer Commanding-in-Chief at RAF Transport Command in 1955 during which period he saw the introduction of the Comet 2 before he retired in 1959. Air Marshal Sir Andrew McKee KCB CBE DSO DFC AFC died on 8 December 1988.

Chapter 4

Getting Frightened

This Wilhelmshaven raid was my first flight on coming back from leave and I was rather glad to have it because the previous night Jerry had been over and kept my wife and me awake with a few bombs not many yards away from home. I thought it was nice to be able to return his visit so quickly. Over the North Sea we ran into thick cloud with a base at three thousand feet. I took the machine up to 11,500 feet and avoided it, just skimming along the top. But that brought us into nastier stuff, electrical disturbances and, what was worse, icing. Blue flames were flickering round the airs crews and the guns. The aircraft was bumping about and it was hopeless trying to use our wireless. It was terribly cold - about minus 29 degrees. At 12,000 feet we got clear of the cloud, except for one great whopper which we had to dodge. Forty miles from our target flak began to stream up from Emden; actually the ground defences were firing at another of our aircraft and we skirted neatly round the barrage and sailed on to our target area. Once there we saw the thick black shapes of buildings standing out clearly against the snow. Some of our boys had already been over and had started five jolly good fires. There was one huge blaze which we reckoned to be about half a mile long and another big circular fire out of which ten explosions came just as I was running over my target. There was no need to bother about dropping any flares; the fires gave us ample light.

I decided not to bomb the biggest fire which was doing very well for itself, so I just waited my time. Then I saw heavy flak being fired at someone on my right. It gave me my chance. While Jerry's attention was distracted we snooped in unobserved. In a straight run we dropped one long stick and that brought the anti-aircraft fire on to us, but by turning round I was able to avoid most of it and swung away to have a good look at the damage down below. Our bombs had set ablaze a large area and we saw nine or ten explosions all of which were quite distinct from the bursts of our own bombs. We toured around for another five minutes though all the time the flak was getting heavier. The extra five minutes were worthwhile for at the end of them one of the fires we had started was running into another and the dark shadows thrown by the buildings on to the snow were being driven away by the light of the flames.There were many other aircraft over Wilhelmshaven and we saw bombs from some of them bursting squarely on decks and dock buildings. Chunks of debris went hundreds of feet into the air and the job done we turned for home.

It was very quiet going home until we got over the North Sea. Then my airspeed indicator got sulky and froze up on me. I thought I would go down to a warmer layer, but the cold clouds were almost down to sea level so I climbed back again to about twelve thousand feet and eventually got over the top and carried on home. There were one or two anxious moments when we were landing, for the airspeed indicator was still frozen up and when you have no idea of your speed it is pretty unpleasant landing in the dark. But we managed it all right; there was a kindly moon to help us.

We all agreed that it had been a very good show.

***Raid On Wilhelmshaven* by 'A Flight Lieutenant', BBC broadcast, January 1941. 'The speaker has been in the RAF for fourteen years. He is thirty and has done 1,900 hours' flying. Since the war began he has been on 27 operational flights, including five over Berlin and one over Turin. On the night of 16/17 January he was the captain of a heavy bomber which took part in the concentrated attack on Wilhelmshaven.'**

'The path from our sleeping quarters ran through the woods and in the early spring the walk was lovely with new flowers appearing each morning, new trees bursting into leaf, the birds and the shy red squirrels. Half-way along, the ground rises a little and so flat is the countryside that it is a surprise to find there is not sea beyond. Instead, below there are more woods with aircraft half-hidden amongst the trees and a little beyond is the wide open space of the airfield. Before half the distance to the squadron buildings has been covered, it becomes obvious that something is brewing. Ground crews are hauling covers off the aircraft and tinkering with engines. Bomb trolleys, coupled together to form long serpents, are rumbling along the concrete paths to the dispersals, the little bays in the woods where the aircraft are standing. The Squadron Commander's car is already outside his office. All these signs point to one thing. Operations are 'on' tonight.'

E W 'Bill' Anderson was a schoolmaster in peacetime and joined the RAF in 1940 as a Pilot Officer in administration. The death of an air crew friend led him to volunteer for flying and, despite 'advancing age' and a suspect right eye, he managed to pass on to training in Canada and then on to Wellington bomber operations on 9 Squadron at Honington. He soon discovered another inherent 'weakness' - he was air-sick every time he flew.

The journey across to Canada had been full of curious and lovely sights and sounds. 'A couple of thousand men singing on the railway platform so that it seemed strange that the searchlights overhead were not keeping time. The little bunch of Poles who manned the ship's after-gun, dark, bitter men who when their country was overrun, had loaded up their boat in America with high explosives and tried to set out on a last voyage to Gdynia. The corvettes battling along through the grey wastes of the Atlantic. The lights shining from the windows of Halifax. The vast railway engines with their faint, plaintive hootings. The oranges, the butter and the sugar. The cold that hit like a hammer. The lovely Canadian girls. New York and skyscrapers as unbelievable as their photographs. And finally Miami and the sodden heat. There ten of us lived and worked with a couple of hundred US Army Air Corps cadets. By day we attended lectures in a co-educational university which reminded us irresistibly of a Hollywood production, with football matches played at night under the glare of arc lamps and beauty contests fought out just as desperately by day. To gain air experience, we used to navigate flying-boats out over the lovely Bahamas, or when at night the black thunder-clouds boiled up over the Gulf Stream, we would fly down the coast of Florida and out into the Gulf of Mexico. And at week-ends we played on the glorious beaches and bathed in the deep blue sea.

'They were wonderful months and yet in spite of it all and at times almost because of it all, we were not happy. We could not help worrying about home. Back in England, we had been full of confidence as to the outcome of the war. Over in America, we saw at last that we had been defeated at Dunkirk but too British to realize it. The Americans never tired of telling how, soon after the tragic evacuation and the fall of France, a short extract of a news-reel had arrived from England which told them why we should never be beaten. It showed British soldiers practicing for the invasion of Germany. To us, fresh from a country united as never before, it had seemed quite logical. To the Americans who thought that we were beaten and were convinced that already we were starving, it was just a magnificent gesture of defiance. And though at first we laughed at their fears, as time went on, we began to wonder. We could see the War now in its true perspective and it was all the more sinister for that. And we longed to get back to be able to satisfy ourselves that there was no truth in what they believed. Slowly their fears were beginning to become ours.

'We finished our navigation training and said good-bye to the States. We took away with us memories of many kindnesses and of a wonderful hospitality, of boys grown up at seventeen and men of sixty still no more grown up. Of girls so slim and fragile that it seemed they might snap if they were touched - and they usually did snap, too! And above all, of charming women, who even as grandmothers never give up the struggle for perpetual youth, fighting for it gallantly to the end with the proud motto, 'We may dye, but we never surrender'.

'So we went back to Canada for our Bombing and Gunnery training. We were sent to an airfield in the middle of the prairies where gophers and coyotes lived and where the yellow grass was alive with grasshoppers. Here we flew in aircraft that had fought all through the retreat to Dunkirk and for their scars we forgave them much. But they certainly were incredibly hot and smelly and there was no means of talking to the pilot, it was a matter of crawling forward and poking a note up to him through a tangle of controls. The 'intercom', the telephone system in the aircraft so that the crew can talk to each other, had been removed. In due course we were ready and one morning we paraded and observer's wings were pinned on to our inflated chests.

'It was a couple of months before I finally set out for home. We crossed on a tramp steamer with Dutch officers and a Chinese crew. The ship wallowed horribly and the decks were cluttered up with crates of all sizes. One night it blew and some of the crates got loose. It was ticklish getting them all secure again, but it was done at last and there was only one casualty. One expectant mother was tipped out of bed, in other words, poor Doris, the ship's pet rabbit, had her hutch overturned. And next morning, the engineer came to us with a long face, 'Ach, the poor Doris, she has had the undercarriage!'

'At last the grey hills of Scotland rose out of the mists ahead and I knew that love of these islands of ours which sounds so stupid if you write about it. The rest of the journey was a rising tide of delight. The English voices on the jetty, the journey southwards by train with the rain falling gently on the little green fields. This wonderful country of mists and soft colours with its little villages and tall churches. And finally my wife Mary.

'After leave I was posted to an Operational Training Unit in Worcestershire. Here at last we flew in operational aircraft with instructors fresh from the real thing. And here I parted with the ten who had been to Miami. We talked vaguely of reunion dinners, but a year later a table for two would have sufficed. And so I was crewed up.

'Fighter boys will assure you that the full joys of flying can only be tasted in fast, single-engined aircraft. Others swear by the thrills of low-level daylight bombing. But those who have been in a heavy bomber crew on night operations will surely count this experience the most precious of all. Seven men, very much alone together in the night, meanness, greed and jealousy swept away by the keen air of danger, are welded by fire into' one. Each loses himself in the crew and so finds himself, curiously secure in spite of danger, comforted in spite of difficulties, somehow safe and at peace in the middle of death and destruction.

'With very few exceptions, the pilot will be the captain. He must be steady, very steady and whatever happens, he may not 'flap' nor get excited. And in the last resort, he must wait in his seat in a blazing aircraft until the rest of the crew are out. He can feel a tug on the controls as the tail gunner leaves his turret and he can count those that clamber past him to the escape hatch ahead. But he cannot always tell if anyone has gone out of the hole amidships. So he hangs on for a little. And if he hangs on too long, his mother will learn three months later that her son who was posted as missing has been killed. It is unlikely that there will be enough confirmation for him to be awarded the Victoria Cross. And owing to the regulations, there is nothing else but a mention in despatches. For no other medal may be given posthumously.

'The W/Op, the wireless operator, is another who may have to stick to his job until the last minute, particularly in the case of having to come down in the sea. Then he stays at his set sending out SOS's so that the crew may have the best chance of being picked up. And for some unknown reason he is often put in charge of the flying-rations. These consist of coffee in thermos flasks, a bar of chocolate, a few sweets, chewing gum and a tin of orange juice. The boys in my crew used to save their orange juice up for Andy Mark III, which was kind of them because it is very refreshing on the way home.

'The two gunners are the eyes of the crew and everything depends on their seeing the Hun in time. It's not an easy job and there's a lot of skill in it. A gunner even has to learn to look not at the enemy' fighter but a little to one side. He can see him better that way. Try it yourself some night and think what it means to do this when that black streak suddenly caught in the corner of the eye may be a Hun-hunting.

'In the early days, heavy bombers carried second pilots, 'second dickies,' as they were called. The idea was not so much as a reserve but rather, to give the newcomer to the squadron a chance to learn his job thoroughly before taking on a crew of his own. As time went on the number of 'second dickey' trips was reduced to a couple, just to give the 'sprog driver,' the inexperienced pilot, an idea of the 'score.'

'The crew was completed by the navigator who was trained to act also as a bomb-aimer. The heavy bomber would therefore normally carry five men, a pilot, two gunners, a WOp and a navigator with sometimes a second pilot coming along as an extra, to pick up 'clues'.

'In 1942, when four-engined aircraft began to come into service, two more men were added, making a total of seven. First, a flight engineer, the pilot's 'mate'. Flight engineers are a curious race. To begin with they are often Scotsmen, like engineers on ships. Secondly, they love engines. Let an engine run with but the faintest trace of roughness and though their lives may depend on leaving the district in a hurry, they will clamour to have it switched off. And if there's a fighter sneaking after you so that the engines have to be run faster than regulation speed, or in RAF language, the motors have to be 'caned', well, it really does hurt the engineer more than it hurts the engines.

'The other addition to the crew was the air bomber. Originally a converted air gunner, he soon appeared as a specially trained crew member. And trained not just to do his stuff over the target, but also to help the navigator by map-reading, or by working the various navigation gadgets on the journey to and from the target.

'Air bombers are curiously detached individuals. Over the target, they lie on their stomachs as contented as chickens, impervious to everything except the target below. At times I think they even forget they are in aircraft. There is the tale of an air bomber who in peace time had been employed by a heavy road transport firm. On his first operation, he did a very meticulous job, directing the pilot very carefully, absolutely according to the book: 'Left-left, steady, steady, right a little, steady, steady, steady, steady, steady - oh, cripes, back a bit, mate!'

'There will always be something between those who have been in a crew together. They become very close to each other. They may hardly talk at all in the air and yet each will be aware of what the others are doing and feeling. The man in my first crew whom I shall always remember was the wireless operator. We called him 'Eddie.' He had a wife in Wellingborough. We never talked much and yet we would have gone anywhere together. Once I was asked to fly with a crew that were rather shaken. They had been shot down into the sea on their previous trip and had paddled their dinghy for hours against wind and tide, only to find when daylight dawned that they were just off shore, still quite close to the aircraft and might have waded all the way. Naturally, they were not exactly the confident and calm collection of men that one would choose as companions when visiting the Third Reich. But Eddie worked his way into the crew for that night, too. He just could not have stayed behind. And that is why, now that he is missing, I cannot bear to hear anyone whistle that silly little tune that was always on his lips.

'I was very lucky in my crew. The skipper was young, nearer to Andy Mark I in age than to me, but he could fly an aircraft as steadily as if it were on rails. And from the front gunner; a cheerful Yorkshire lad, to the tail gunner, dark, silent and deadly efficient, they knew their jobs thoroughly. The only exception was myself and I was perhaps not entirely to blame. It was before the days of bomb aimers and flight engineers and in addition I had to keep a record of the weather, take photographs, adjust the compasses, note down intelligence gen and so on. I was jack-of-all-trades and master of none. Officially I was known as the navigator.'

'The first job of the day is the NFT, the night flying test, a short flight to check engines, bombsight, guns, intercom, lights, oxygen and so on to make sure everything will be on top line for the night. We stroll over to our aircraft. On the side of the fuselage by the red, white and blue roundel, two letters for the squadron are painted and a third for the particular machine. Ours is Q, or as we call her, 'Q-Queenie'. Aircraft are always referred to in this fashion to avoid muddles; for instance, Q and U sound very much alike when spoken over a crackling wireless, but 'Queenie' and 'Uncle' are quite unmistakable. There is an official list of words for the letters and a good many unofficial words, some of the more printable and subtle being 'T for Two', 'L for Leather', 'R fer Mo', 'Y for Nagging' and 'V f' la France'. The ground crew were busy on 'Q-Queenie', doing something to one of the engines. These RAF ground crew may not be very impressive to look at. They are dirty and unkempt, for aircraft are messy, oily things. Their pockets bulge with spare parts and weird tools stick out of their uniforms in unexpected places. And their language is a wonderful thing to listen to. Yet though they are not and in fact cannot be spit-and-polish men, yet, they represent one of the greatest miracles of the war-time RAF.

'Only a few months ago all but one had been civilians who had never seen the inside of an aircraft. Yet here they were responsible for the day-to-day upkeep of an incredibly complicated modern bomber. And so good was their work that night after night the bomber would go out and the question of breakdown, apart from the effects of flak or cannon or icing, just was not considered by those that flew. For example, a trip to Milan that involved lifting a staggering load of bombs over the Alps was always considered a 'gift', a 'piece of cake'. The danger from flak or fighters was negligible and the danger of failing to return for mechanical reasons just never entered into the calculations. And that is a measure of the confidence that air crew placed in the faithful care of the ground crew.

'One of the secrets of this success, though the dirty mechanics would look at you pretty blankly if you suggested it, was pride. The ground crews were proud of their job, proud of the fact that the machine they were looking after might be going to Berlin that night to help smash Hitler. So they treated their aircraft as if she were their own and indeed, in their heart of hearts I think they believed that she was. At least, she was a concrete expression of their hatred of the Hun. So they tended her with loving care. They would patch the bullet holes with pride. They would paint up a bomb on the side of the fuselage to mark each operation. And they would watch anxiously how the crew to whom their aircraft was loaned for the night handled their child. And a heavy landing jarred them even more than the people inside and their faces in dispersal were heavy with reproach. Sometimes their aircraft failed to return after an operation. They would be sorry about the crew, but this happened so often that being posted as missing began to mean something rather like being posted to another station, for the crew were so much alive when they saw them last that it just was not possible for them to be dead. But they would feel badly the loss of their aircraft, into which they had put so much. Though being men and so being faithless, the next Q-Queenie would quickly wipe out the memory.

'Soon the ground crew has finished, the aircraft is taxied out and we take off and climb over the wood at the end of the runway. We circle once and set off for the range a few miles away to test our guns. Everyone is busy checking the gear, the oxygen, the intercom, the lights and so on. Soon we are over the range, a couple of steep turns, the weak clatter of guns felt rather than heard above the roar of the engines and then back to land and to report everything correct.

'We go to lunch and there are pills on the table in little saucers. We take a couple, not because we care whether they contain vitamins or not, but just because we know it pleases 'Doc.'

'Lunch is a simple affair. Apart from the glass of milk which all air crew get, we have the same rations as the men and we serve ourselves from a large hot-plate. Looking round the room one is struck by the fact that the youngest faces are those of the flight lieutenants and squadron leaders, for they are the regulars who joined the Air Force straight from school just before the war. The rest of the air crew are generally older, while the oldest men in the room, those with last war medals, seem invariably to be pilot officers (probably on probation, too!). The talk is of many things, chiefly frivolous. Of last night's 'party' and the sheep that was put to bed with old 'so-and-so' who was 'out to the world.' And how odd it looked with its head on the pillow and its eyes staring glassily up at the ceiling. Of 'Joe's' awful efforts at 'circuits and bumps' (practice landings). Of the CO's dog and its shameless behaviour. Of 'Jane', In fact, of everything except the one thing that is on everybody's mind-what is the target for tonight?

'After lunch we go to Station Headquarters. We crowd up the stairs to the briefing room and conversation 'stops. For on the wall on the left-hand side of the long room is a map with the route marked in coloured tape. One glance is enough to see the target, whether it is a 'hot' one or a 'piece of cake'. If it's a 'piece of cake', we can all relax. If it's 'hot', everyone acts according to his type. Some nib their hands with glee, they are the very nervous ones; others whistle in dismay, exclaiming loudly that they are breaking down under the strain of operations. These latter are the sounder: there is a third type who wear a mask that tells nothing and amongst these you will generally find the heroes and the cowards. For the hero is generally the coward who has overcome himself, so that the very best and the very worst have a great deal in common.

'The target for that first operation was an easy one. We all sat down and the CO stood up with a long slip of paper in his hand and began to read in a dull, rather monotonous tone as if he were just a little bored with the whole business. It sounded odd coming from such an impressive figure, six feet of toughness with a shock of black hair and a pugnacious chin that earned him the nickname of the 'Hammer of the Hun.'

'Briefing is a disappointing ceremony. One feels that the Air Forces of other nations must do it much better than we do. The Germans undoubtedly sat erect in rigid rows, rising at regular intervals to click their heels and cry 'Heil Hitler'. But the stupid English sit on hard chairs and look a little bored while one of their number reads out a list of instructions in a rather furtive fashion. This is not because the English have no feelings, 'but rather that they are afraid of them. Luckily, briefing is usually brief.

'The next stage is 'flight planning'. We mark our proposed route on our maps and charts, measure distances and calculate how long each stage of the flight will last. The idea is partly to act as a reliable basis for the navigator to check his work in the air and partly as a guide for the pilot in case the navigator should get killed. And when 'flight planning' is over, we go back to the mess.

'The time until take-off is always trying. First of all, the grass looks very green and smells very good and the world seems to be a very pleasant place. Secondly, you do not feel quite yourself. Any other night, of course, you would have been as keen as mustard, but to-night of all nights you feel just a little under the weather. Secretly and privately somewhere inside is a Hyde that hopes that the operation will be cancelled, or 'scrubbed' as the expression is. Though if the operation was to be 'scrubbed', immediately the Jekyll would rise up and be indignant and disappointed, so particularly anxious to fly - another ruddy ' scrub', oh hell! And so between the two squabbling inside you, the time passes slowly.

'There was a Nazi film showing British 'Air Huns' just before take-off. They were lolling about with long drooping faces and long drooping mustachios, like a lot of unhappy prawns, guzzling down vast quantities of alcoholic liquor to sustain their failing courage. I don't know whether the Germans got this idea from the Luftwaffe, but if they did, it explains completely why the German bombers so often did no more than 'plough the fields and scatter'. Nobody would be such a fool as to drink before operations, for the job needs cool judgment.

'But stay, what is this? Just before take-off, I shall cautiously extract a pill from a screw of paper and furtively gulp it down. Drugs! Yes, I must admit it. It is the same sort of drug that my grandmother used to take before going on the paddle steamer to Boulogne. For I suffer from air sickness. So very trying! I used to have to carry a brown paper bag with me, but these pills solved the problem. And the only other drug that is used is caffeine which some gunners take to help keep awake!

'So we sit about the mess, turning over the pages of illustrated weeklies with their pictures of weddings and beautiful debutantes who sometimes do war work but never do piece-work. Or perhaps we play a game of shove-ha'penny with half a mind on the weather outside and the German guns. Supper is eaten rather deliberately. Then comes the time when a little sleep before take-off seems indicated.

'I go to my room and in rather a shamefaced way I carry out a little ceremony. I write to Mary and leave the letter rather obviously on my shirts just inside my top drawer. If I don't come back I know someone will find it there and send it on for me and I hope Mary will understand. Then, as it is still a long time before take-off, I get into bed and lie looking up at the ceiling.

'Soon after midnight, the van arrives and takes us down, to the crew room.

Everyone is busy putting on flying-kit, swaggering a little and laughing rather easily. The atmosphere is exactly that of a changing-room before a football match. Finally we are all dressed up and stroll out to the waiting lorries. The gunners waddle phlegmatically in their layers and layers of clothing. The pilots are slim and slick with mustachios bristling over gaily-coloured scarves. The navigators, very earnest and rather harassed, are bowed down with bags and instruments and mysterious boxes and are followed by their faithful satellites, the wireless operators, picking up the bits and pieces that fall out of the overfilled satchels. The lorry is waiting and we climb in and then trundle off round the perimeter track to the waiting aircraft.

'As we arrive at each dispersal, six or seven clamber out. 'Cheer-oh,' 'have a good trip,' 'see you in the morning.' Then the lorry moves on. Soon it is our turn and we get out.

'There stands 'Q-Queenie', black against the starlight and round her the ground crew to wish us good luck. We go underneath her belly and look up at the bombs hanging between the open doors and it does not seem possible that they can kill. Then while the gunners have that last cigarette and George the skipper has a look round the outside, I climb up into the aircraft, Eddie handing my gear up to me, my maps, chart, instruments, sextant and, so on and the next five minutes is spent stowing everything away and checking that nothing has been forgotten. We hear other aircraft starting up and the skipper climbs in. There is a clatter as each engine is started and a few minutes later a roar as each is raced as a final test. Finally George calls over the intercom: 'All OK?'

'OK Skipper' comes back from each of us in turn and then a shaking tells me that we have begun to taxi to the place where we shall take off. The engine noise rises and falls as we bowl along over the' ground and then dies away as 'Q-Queenie' swings round and stops at the end of the runway.

'The take-off will always be to me the dramatic moment of an operation. I am sitting in a brightly-lit cabin, well blacked out. In front of me is a sort of desk, the chart table and on it are maps, chart pinned down with drawing-pins, instruments, pencils and India-rubbers stowed in the lid of a box that held a last year's Christmas present. Pinned on to the woodwork on my left is a photograph of Mary with Andy Mark III in her arms. On the right is a bar of chocolate, part of the special flying-rations that we are given. Farther to the right is the back of Eddie's head; he is leaning over his wireless, fiddling with something. Just past him is a door and outside is George in the pilot's seat, with Tommy, the second pilot, standing beside him - for those were the days when we carried second pilots. As soon as we are near the target, I shall have to open that door and crawl forward past George to get to the bomb-aiming position, but at present I can sit here warm and comfortable, shut in on the other side by a door which leads aft down the fuselage to Harry in the tail.

'Ahead of us stretches the flarepath, two rows of twinkling lights. At this moment, George will be watching the little black shape which in daylight is a van rather like a bathing-machine. The moment he sees a green light shine out from that black shape, he will open the throttles and take off the brakes and the engine noises will swell into a shattering roar as we roll off down the runway into the night - and to Germany.

'Someone says over the intercom, 'They've given us a green'. The noise increases steadily, the aircraft begins to shake and tremble and we are off. The noise and the shaking increases and the aircraft begins to bounce uncertainly. The roar continues and with it the knowledge that the black woods at the end of the flare-path are rushing towards us. Suddenly comes the awful realization that something has gone wrong. The aircraft cannot get off the ground, it is 'flat out' and yet it cannot make it and there is a load of bombs on board. What has happened - she just will not come unstuck, what the hell has happened - what the -----. The bouncing stops and feeling a little sheepish, I pick up my pencil and write in my log '0207, airborne BASE.'

'And now it is the job and there is not time to think. We set course and are on our way and I check and cross-check our position as we go along. After about a quarter of an hour, Cyril from the front turret rings up Harry at the back: 'Hallo, Harry. How-yre doing, lad?' 'OK Cyril, you OK?' 'OK Harry.' And then they don't talk any more. This little conversation is a sort of ritual and I don't think they would be happy without it. For a gunner's job is a lonely job. He just has to sit and watch, peer out into the darkness and wait for it.

'Soon we are coming up to the English coast and I go forward. Dimly I can make out a dark smudge on the ground below and I know that the dark smudge is a town. It is a peculiar feeling looking down on a town as you flyover it on your way to a target. There are people in that town below just like yourself, sleeping in their beds or turning over uneasily as you go overhead.

'The English coast appears with its thin line of white surf breaking on the seashore. Then back inside to check my work and to calculate the time at which we ought to arrive at the target. About fifteen minutes before we are due, I unpin the picture of Mary and Mark III and stuff it into my Mae West. Then I clamber forward, past Eddie who nods - he is sitting with a 'thriller' in his lap so I can only suppose that he does not find operating exciting enough - past George who gives me the thumbs-up sign and so to lie on my face and peer out through the perspex bomb-aimer's window.

'We have just crossed the enemy coast. It seems surprisingly like our own and is very quiet and still. I look forward and there is what must be the target. A few fingers of white are waving gently to and fro. Suddenly they begin to join up and form a cone and at the same time, little threads of red and green light begin to snake slowly up from the ground below. Somewhere ahead is another aircraft and the searchlights have found him. Little pinpricks of reddish light begin to stab at the head of the cone-shells bursting-so someone is getting a welcome. After a while the searchlights begin to straggle off in ones and twos and finally they are all waving about fretfully trying to catch their victim again.

'Soon the fingers of light become stronger and as we come up to the target, one that was feeling for us shines for a moment in my eyes with a bluish light. Spurts of light begin to appear on the ground as the guns open fire. Ahead is the harbour and I can make out the edge of the docks where the moon is shining on the water in the bay. Then flashes begin to appear in the sky around us and once we bump a little as we fly through the air disturbed by a burst and we can smell the cordite. But it is all very unreal, for we can hear nothing above the roar of the engines which has drummed for so long in our ears that we are no longer aware of it. And so the effect is of absolute silence.

'I am lying down fascinated by the weirdness of the sight and watching the target slowly creeping up towards the point where I shall let the bombs go. As soon as it reaches the two little luminous points, I shall press the 'tit', the little push button that I am holding poised in my right hand and the first four bombs will fall. 'Left-left a little-steady-steady-steady' and then I suddenly realize that I have forgotten to set up the sight properly.

'We go round again while I curse myself for my carelessness and set up the sight again and then we make two runs. As the first four bombs go, I crane forward and watch them in the moonlight. They seem to float down gently underneath, then suddenly as the ground below comes into focus; they shoot forward towards the docks. Soon they are lost and I wait. A sudden flash of light from somewhere behind tells us that the flashlight has worked and so we ought to get a picture, but still no flashes that look as if they were our bombs. Frantically I check, for there is a switch which if it is left up operates a safety device and the bombs will not explode when they hit the ground. No, everything is as it should be, as far as I can tell I have not made an idiot of myself a second time, the bombs were properly fused. So we make another run and this time we see four flashes in a sudden succession close together on the edge of the docks. Then we turn for home. Slowly, very, very slowly, one wing dips down amongst the searchlights while the other reaches up towards the stars. For a long time we seem to hang motionless above the target, then gradually we level out and slowly the searchlights are left behind as we head back to England.

'Dawn is breaking as we come back over base and the lights from the flare-path shine palely from the ground below. George is calling base. 'Hallo, Anvil, hallo, Anvil. Postcard Q-Queenie calling, Postcard Q-Queenie calling. May I land, please? Over.' Back comes a WAAF voice from the control tower, 'Hallo, Q-Queenie. Anvil answering. Yes, you may land now, you may land now. Over.'

'OK. Q-Queenie coming in to land' - and then follows the usual routine, 'Wheels down' -' Flaps down' - George and Tommy are checking that everything is ready for landing. The engine noise rises to a whine as we come in to land and the high note thrums anxiously through the fuselage. The engines splutter and choke, the whole aircraft slowly tilts up as George eases her in, there is a gentle bumping and a cheerful comment from Cyril, 'Just like our drawing-room ----- all in it!' and the operation is over.

'We taxi to dispersal, the engines are stopped and the silence deafens us. The ground crew come up grinning and we clamber out and stretch ourselves. The lorry is waiting once more and we rattle off to interrogation. Up those stairs in Station Headquarters again, but this time with no forebodings. Waiting for us are cups of hot tea and, if we want it, a tot of rum and sandwiches and cigarettes dispensed by WAAFs who are kind and understand that we are too tired and deaf and dirty to be gallant. We cluster together in groups or loll in armchairs waiting our turn to be called next door where, over more tea and cigarettes, we will tell our story to the Intelligence 'types'. As this is our first trip, we shall talk a lot, but the Intelligence Officer will listen to us patiently for he understands. After a few trips we shall be less voluble and less positive of what we have seen. And after twenty or thirty, we shall say little, but what we do say will be 'pukka gen' and stuff that he can rely on.

'After interrogation we go back to the mess, not to bed at once, for there is a breakfast of bacon and eggs and everyone is talking and nobody listening. After an hour or so, we who had screwed ourselves up have unwound. And so to bed and strangely soon to sleep.

'A few hours later, we were woken up again and told that we were 'on' again that evening. We went to briefing and the target was another easy one. Afterwards they showed us the photograph taken the night before with our first four bombs. Technically it was a masterpiece, the docks and the buildings in the town shown up as if it had been daylight. And what was more remarkable, we had achieved the rare feat of actually photographing our bomb bursts. What was not quite so gratifying was that there on the picture, plain for all to see, was the explanation as to why we had seen no flashes. For there were those first four bombs, four large splashes right across the water of a little inner harbour!

'We took off that evening just before sunset. The fields below looked very green and peaceful, with the trees painting long shadows across them. Here and there a thin line of smoke was curling up from a village and one could picture the folk sitting round their fires reading the evening papers. Slowly the sun sank down through the pink-tipped cloud-tops in the west and darkness began to flood the earth below while the aircraft was still bathed in golden light. For in the air, night does not fall. It rises stealthily out of the earth beneath till finally it reaches the sky above.

'We chattered cheerfully amongst ourselves like small boys playing at soldiers when there is no danger. For were we not an experienced crew? Had we not done one whole operation? Flak! Pooh, it was harmless, very pretty to look at but nothing like as dangerous as we had been led to believe. Fighters? We had never even seen one and we had done one whole trip!

'Slowly the light faded and as we flew out over the sea the stars began to peep out. I was sitting at my table busy checking the effect of the wind and everything was going according to plan. Suddenly a clatter broke out as the tail turret opened fire and we shuddered. Harry's voice came through quite quietly; the enemy fighter had broken away to starboard and was lost in the gloom. And now our aircraft would be between the fighter and the afterglow from the sun, silhouetted.

'And then I knew fear. Somewhere outside there was someone crouching behind guns circling round to come in again and kill me. I curled myself up a little tighter in my seat and waited. He did not come in again; perhaps Harry's burst had gone home. But for the rest of the time that I flew over Germany, I was still curled up and waiting. For in that brief moment when I watched a pattern of holes appearing in the fuselage beside me I discovered that I was a coward, hopelessly and horribly afraid of fighters. - Flak and in particular heavy flak, is somehow impersonal. It is fired at the aircraft and not at you. I always imagined a lot of earnest little Teutons wearing spectacles, busily working out exactly where you are, loosing off a shell hopefully and then - watching-watching-watching -' Donnerwetter, missed him again!' But fighters are different. They are personal, they are out to kill, they are aiming not at your aircraft; but at you.

'We went on to the target, but it was misty and we could not find it. Searchlights came feeling for us and pointed towards us along the ground, while others flashed straight on to us and just as mysteriously went out. Eventually we came across a Hun aerodrome which fired a green cartridge at us to encourage us to land. So hoping that our acquaintance of earlier in the evening would be landing there, we let them have our bombs instead and all the lights went out very quickly. Then we came home. We landed and found the wing badly damaged and blessed our self-sealing petrol tanks.

'Poor old 'Q-Queenie' went into dock and a few nights later we went to Essen in 'F-Freddie'. Essen was the hottest spot in what was then the hottest part of Germany, the Ruhr Valley, known to the boys sardonically as Happy Valley. We did not talk much on the way to the target, for too many of our friends had failed to return from trips to Krupps' works.

I comforted myself by calculating the chances of getting shot down or ' going for a Burton' as the RAF say, divided the number of years that I might reasonably expect to live by this figure and metaphorically presented my country with two years of life. It sounded so much better than getting killed and mathematically, at least, it came to the same thing! The reasoning works this way. In those bad days, the chances of not coming back were about one in twenty. I could expect another forty-odd years of life, as I came from a reasonably long-lived stock. Twenty into forty goes twice, - so that in effect I was not offering my life, but merely a paltry two years, that is a proportion of it depending on the chances of my getting shot down. I found a lot of comfort in this calculation, though looked at in another way it did not seem too likely that I would be able to continue making this offer during the whole of my first tour!

'We arrived over Essen three minutes too early and discussed the question of circling so as to bomb at the right time. This was our first trip to the Happy Valley and things were very quiet, so that we realized that the danger of the Ruhr defences had been greatly exaggerated. There was absolutely nothing to worry about. We started a turn to the left and as we did so, we noticed a good many flashes from the ground.

'There were several hundred guns in Essen and each battery was fitted with radiolocation. For the past few minutes, the gunners must have been plotting us innocents pretty carefully. We had been flying dead straight and level and at exactly the right height. They waited for us, got us taped and then let us have it. One moment we were as bold as brass hats; the next, our hearts were in our oxygen masks.

'In a few moments the air was filled with bright flashes and we heard a new sound, the crunching of near bursts and the aircraft shook and trembled and splinters rattled against the fuselage. Then there was a blinding flash and a crack and F-Freddie leapt and shuddered and then began to tremble violently. We had been hit properly this time.

'We did not know what had happened. All we could tell was that the aircraft was vibrating badly. We dropped our bombs and shuddered off home. The vibration seemed to come from the starboard engine, but when we turned it off, we lost height, so that we had to keep it going slowly. I still have the log written on that trip and it is as if a man with the palsy had made the entries on the way home. We limped back across the sea and at last made base, landed on one engine and had a look at the aircraft. The wing was full of holes, the tip of one propeller blade had been knocked off and the consequent vibration had all but pulled the engine out of its mountings and the main spar supporting the wing was cracked.

'The next night we went to Essen again, this time in 'C-Charlie'. She was a good aircraft, normally flown by a dour New Zealander by the name of Bill. Before take-off, Bill came fussing around and begged us to take the greatest care of 'his' precious aircraft. We made all sorts of promises and set out intending to be scrupulously careful. The road to hell is paved with good intentions!

'It was a nightmare of a trip, for this time we were caught by searchlights. We threw out the odd beer bottle which makes a scream just like a bomb as it falls, the idea being that this would discourage the types below and they would dowse their lanterns. What a hope!

'Everybody who flew on operations hated searchlights. Partly because of the fighters that would come in when you were dazzled and partly because you became at once a target for all the flak in the parish. But worst of all was the horrible feeling of nakedness that it gave and you longed to escape into the friendly darkness.

'The old hand would seldom get caught and it was fascinating to stand by and watch him play those long, white fingers. As one came feeling towards him he would dip a wing and deliberately swing the aircraft right into the beam so that it grew larger and then suddenly flicked past and was gone. Then it would waver about for a little in a fretful way until it had decided where you had gone and back it would come. And once more the old hand would turn in towards it. If he were to turn away, the searchlight would gradually overtake him and pass across slowly enough for the man on the ground to spot the aircraft and hold it. And then the other beams like the limbs of some giant white octopus would come sweeping in from all around and the victim would be caught and held. And it was no good twisting and turning wildly in an attempt to dodge. The wisest thing was to put the nose of the aircraft down and leave the district hurriedly.

'But George was not an old hand and what was worse; he became dazzled and threw the aircraft about aimlessly. His violent corkscrews were but vicious circles for we got lower and lower as we twisted and turned and they hosepiped us with light flak. We were continually hit by small splinters and finally Eddie got one in the foot. The guns promptly decided that honour was now satisfied and left us alone. And the peace that filled our souls was like the peace that comes when the dentist lays down his drill and picks up the pestle and mortar.

'So for the third time running we came back with a battered aircraft. Bill, the 'owner,' was pretty outspoken about it. 'I lent you a perfectly good aircraft and you silly --s bring back a -- colander.'

'A few days later we were sent on leave. And there amongst the Devon lanes it seemed that the war was all a dream! I began almost to wonder if it could be true, until I went with Mary and the boys for a ride in a motor-coach. I had to sell the children the idea of coming back by train. I could not bear to ride any more in a bus. The bumping noises below took me back to those nights over Essen with the shells bursting under our belly.

'The battle of the Ruhr in 1942 cost many good crews. The squadron lost steadily like most of the squadrons nearby and the faces in the mess were continually changed. It was a stroke of genius on the part of the authorities to give us, one night, a complete change of target, a complete change of method and a complete change of tactics. One afternoon we went to briefing and found that the target was a new one, nobody had been there before and the plan was simple and full of possibilities. The first arrivals were to drop flares, then aircraft carrying incendiaries were to get the heart of the target well alight and finally high explosive would complete the destruction. The name of the target was Lübeck.

'This raid will always stick in my mind. To begin with, it was perhaps the first really scientific attempt to blot out an industrial target covering a large area. It was a finger pointing to what was to come and it was brilliantly successful. The night was clear with thousands of stars above and no cloud below to cover the nakedness of the land. We had to keep a careful watch for fighters. Just as we flew past the island fortress of Heligoland, we saw tracer in the sky and an aircraft caught fire and fell blazing into the sea and burnt for a while on the water. Soon we had crossed the enemy coast and passed safely through the Kiel Canal defences. Ahead of us lay the target, but we were early, so we flew into the Baltic to use up the few minutes we had to spare. It was frozen over and there were lines along it as if the Germans had been having skating races. Then we turned and ran into the target.

'As we came in, a few flares went down in front of us and we saw the target and let our incendiaries go. We were not quite the first, for as I was watching them falling below like a shower of little silvery fish, a load went down, a smear of twinkling lights. A few moments later and our 'stick' splashed across the target and there blazing on the ground was a vivid white V for Victory and then loads began to fall thick and fast. Five minutes later when we set course for home, the centre of the target was an inferno, deep, glowing, red fires were burning and we all felt some of the horrors that must have been going on down there.

'Back at base we discovered that we must have flown alongside one of the other crews in the squadron all the way, though we had not seen them. For the aircraft we had seen falling in flames by Heligoland was a fighter that they had shot down into the North Sea on the way out and it was their stick which had first gone down across the target.

'The operation was a wonderful tonic to the squadron and to our crew, George had been badly shaken by those nights over Essen, but he would never admit it. We could tell that his nerves were affected by the way he would fuss over little things going wrong and by his vociferous keenness to operate, as if he had to convince even himself. But this latest show, with the certainty that for once we had given the Hun far more than we had taken, would surely enable him to turn the corner.'

Following the attack by 234 RAF bombers, mostly carrying incendiaries, on the old Hanseatic City of Lübeck on the night of 28/29 March when about half the city, some 200 acres, was obliterated, Hitler ordered a series of Terrorangriff (terror attacks) mainly against English cities of historic or aesthetic importance, but little strategic value in reprisal for RAF Bomber Command raids on German cities. On 24 April Baron Gustav Braun von Sturm, Deputy of the German Foreign Office Foreign Press Department, said, 'We shall go out and bomb every building in Britain marked with three stars in the Baedeker Guide [sic - Karl Baedeker's series of guide books never marked any buildings with more than two stars!]'. Exeter in Devon was bombed on 23 April and was followed by Bath in Somerset.

The youngest of three girls, Myfanwy Hams lived with her family in Exeter. Her father was a local doctor and a Senior ARP Warden. 'I remember being woken up and wasn't dressed. There was no time to get dressed because the sirens were coming and the bombers were coming in straight away. My mother and I were dressed in our nightdress and dressing gowns throughout the whole thing. I said goodbye to my father and went in to the shelter. He went off and my mother and myself and the dog, we shut ourselves in to that cage [the Morrison Shelter] and the noise was horrendous but again we were used to it, whining and screaming. The drawing room was lit up. You could read because of the flames which my mother realised were in the garden. So she went out to see if she could put the flames out in the garden and then she looked up and she realised the house was burning. So she came back and said we have got to get out and I knew that if we got out we wouldn't go back and I begged her to stay so to keep me happy she stayed. It couldn't have been more than a couple of minutes and then she said, come on, we have got to go. I was on her right and she had the dog in her arms. We walked from the drawing room through into what was called a lounge which was huge, right out into the hall and the stair case on the left was burning as we walked. She left the front door open and shouted to the cat, there was no chance of looking for the cat. We crossed the drive and there was a shrubbery opposite the drive and we lay there as we had been told, you had to lie flat but your chest had to be slightly off the ground. You shouldn't be completely flat. Then we were dive bombed. A couple of times people tried to rescue us. The air raid wardens tried but they couldn't hear our shouts because of the noise of the raid. Eventually, one of the wardens came to rescue us and told us to run. I stopped to look back to see a huge sheet of flames come out of one of the attic windows. Then I knew that the house was gone.

'My father had been told of a report that our house was burning and my mother and I hadn't been found. He got in his car and put his foot down. He stopped to drag two bodies to the side of the road and to confirm that they were dead. Then he had to stop again to beat out the flames that had fallen through the car and were on the back seat. When he got home he could see the house had gone. He went into the garden and opened his mouth to call for us and he lost his voice. He couldn't shout; he couldn't do anything. When he found us next door, I greeted him with what later he thought terribly funny. I said, 'Hello daddy. We have been blitzed.'

'The next morning my mother and I sat and watched our house disintegrate. It was still smouldering as we watched the dawn come and we heard the most glorious dawn chorus. It was a perfect day. It was a beautiful May that year and we had a liliac in full bloom and I can remember the brilliant blue of the liliac, the sky gradually getting blue, the sun mounting, the birds singing and our house collapsing.'

Michael Lee, a schoolboy living in Bath never did hear separately the bombs which so narrowly missed his family and destroyed two pairs of houses immediately behind them and in the process tore apart their house. 'The front door was blown in and the kitchen door was torn off its hinges (just outside our cupboard) and pulled to the bottom of the kitchen - towards the explosion - by the inrush of air into the vacuum created. The back wall of my bedroom was bowed into a semi circle - again towards the explosion - and the whole roof of slates was lifted and dropped back in disarray, followed by a great deluge of debris. However we were unharmed! In the morning, when it was finally quiet and daylight arrived, Dad went to get our Austin 8 from the Chaucer Road Garages (he had a tiny ration of petrol for Home Guard duties) and I wandered round and up Milton Avenue to look at the back. The pair of houses immediately behind us was a heap of rubble and swathes of timber, most of it cascaded into our back garden. As we set off, early that Sunday morning, I saw that there were houses missing in Kipling Avenue, the 2nd to the 4th down from Chaucer Road on the city side, which were just a heap of smoking rubble.

'The Blitz may have lasted only two full nights, from the evening of Saturday 25 to the morning of Monday 27 April 1942, but those moonlit nights were a defining moment for us. We had lost the innocence of childhood and had been thrust into full awareness of life and death, of terror and distrust and the ugliness of destruction and hate. And yet we learnt how to care for each other and how to pull together in a unison and harmony that transcended the pre-war boundaries of society.'

Norwich, a city of 126,000 citizens, was the third target in the 'Baedeker' series of raids that would kill a total of 1,637 people, injure 1,760 and destroy 50,000 buildings.

At 2015 hours on the night of Monday 27/28 April 28 German raiders were identified on radar heading for Norwich and twenty Ju 88s of Kampfgeschwader (KG) 30 laid mines off the coast. Only three AI (Airborne Intercept) equipped Mosquito night-fighters, nine Beaufighters and ten Spitfires were available to intercept and the city was not protected by enough anti aircraft batteries. Though AI radar contacts were made none of the raiders were shot down. This was the first raid by I/KG2 'Holzhammer' since converting to the Do 217E. Together with IV/KG30 and II/KG40 from 23.40 to 0045 hours they dropped 185 HE bombs and incendiaries weighing 50 tons on Norwich, killing 162 people and injuring 600 more. Fourteen of the victims died instantly when their shelter near the 'Dolphin' public house took a direct hit. Thousands of buildings were damaged and 84 people were dug out of the rubble alive. Some reports said that the enemy bombers machine-gunned the streets.

Teenager, Nora Norgate, was one of thousands of Norwich citizens who in spring 1942 was forced to endure frightful German bombing and deprivation, would never forget that night.

'The air-raid sirens began their mournful wailing at about 11 pm. We stirred in our beds waiting for the distinctive sound of the hooter, which told us enemy bombers were getting closer. It went almost immediately as we scrambled from our beds, hurriedly dressed, grabbed our torches and had just began to race downstairs and out to the Anderson Shelter in our backyard when the first of many bombs came whistling down. We cowered helplessly on the stairwell in the middle of our house hearing the frightening whine of falling bombs, the awful droning of the enemy planes and the house-shaking explosions. Then the windows suddenly shattered and were blown inwards, closed doors were blasted open, ceilings cracked, then collapsed in clouds of choking dust around us. We were absolutely terrified and were convinced we wouldn't live to see the morning. My sister, aunt and I clung closely together while wondering if my younger brother, who was at that time a messenger for the Air Training Corps, was in the immediate area. It later transpired that he was quite safe. Flares lit up the whole city like daylight as they floated down from the stream of bombers, dozens of swaying searchlight beams, streams of bright tracer bullets were flying, anti-aircraft guns booming defiance. I can't remember how long it all lasted, but it seemed forever.

'When we could no longer hear the bombers overhead we ventured out into the street. Shocked, shaken and in tears, we saw an unbelievable scene of destruction. Most of the houses in Belvoir Street were damaged. A few had been reduced to scattered piles of fiercely burning matchwood and rubble. Many people had been killed, even more injured, some seriously. Other areas of the city had been hit much harder. By some miracle Number 31 appeared to have been one of the least damaged houses in the street, but it would be some time before we got our windows replaced and all the repairs completed. The sound of the anti-aircraft gunfire, the strict 'black-out,' the Air Raid Precaution warden, the food and clothing ration books issued to each family, gas-mask drills, steel helmets at work, sleepless nights in the shelters, shortages and long queues for everything and evacuation drills, were a part of every-day life.'

Ted Harvey, an NFS section leader, recalled:

'Parachute flares lit up the city and almost immediately this was followed by the series of explosions as each 'stick' of bombs took effect. Mixed with the high explosive bombs were the incendiaries and in no time at all the glow of large fires became visible all over the city. The divisional fire control at Bethel Street was immediately flooded with calls from police, wardens and citizens of the city, nearly 200 calls being received within minutes of the initial attack. The control staff - firemen and firewomen - went into action. Their hours of training had now become a reality. They ordered pumps and appliances from the sub-stations and soon these were racing through the streets of Norwich. But fire calls were being received in stations as fast as telephone lines could carry them and more fires were burning than there were pumps and crews in the city. Reinforcements were ordered by the divisional mobilising officer and appliances began to arrive from all quarters of the region and eventually from as far away as the City of London itself. The local crews worked like demons, pumping from water mains where these were intact. Where they were not, they got to work from steel and canvas dams from flooded bomb craters and from the river through miles of canvas hoses. So busy were the firemen many failed to hear the sirens sounding the 'All Clear'. They worked all through the long night and well into the next day until relief crews from long distances took over.'

Two 500kg HE bombs fell on Chapel Field Gardens (fortunately one was unexploded) causing many casualties 'Ossy' Osbiston, a member of the rescue service, recalled: 'At 11.15 enemy aircraft flew on over the coast and the first plane dropped chandeliers over the centre of the city and then at short intervals explosives were dropped at random. I had to ask for volunteers to come with me to the worst incident at Chapel Field Gardens. The site was a bad one. Young and old were entombed in an underground shelter. Two sisters, about 15 or 16 were dead. An elderly man was in a sitting position with smashed legs. A very beautiful young girl was suffocated, but she was lying on top of an old lady and this had saved her life. Part of a concrete reinforcement had gone through her hand. I cut the iron off and my First Aid man covered it and she was sent to hospital. She recovered, as did the chap with smashed legs.'

The RAF's Air Warfare Analysis Section would later conclude that the total weight of High Explosive (HE) in this first raid that hit the Norwich Civil Defence region was 45 tonnes of bombs, 96% of the total dropped. Initial casualty figures of 53 killed were later corrected to 162 killed and 600 injured, making this the worst loss of life in a raid on an East Anglian target in WW2. City Police estimated that 7,000 houses were damaged. The Ministry of Home Security reported that Extensive damage was caused, 'mainly to the poorer residential quarter' and that 'considerable damage was done to Public Utilities'.

'Two nights later [on Wednesday 29/30 April] Norwich was again the target for German bombers' continues Nora Norgate. 'More high explosive and incendiary bombs fell, causing more fires and more loss of life but we were in the comparative safety of our Anderson shelter that night. [36] Many people were so apprehensive that they left the city during the nights, sleeping in any kind of shelter available in the surrounding countryside and returned to their homes the following morning. During the next seven or eight days after those two air raids, our family would leave our home after tea, walk out of the city, carrying blankets, pillows, sandwiches, hot tea in flasks and our torches to the Mile Cross Bridge and sleep under the bridge each night. We were up early the following morning, walked home and then went to our various places of work. About 30 barrage balloons were installed around Norwich shortly after the two raids and in early May 1942 they proved their worth because here was another, larger air raid, but the bombs fell on the outskirts of the city.'[37]

While Norwich reeled from the constant bombing at least its good citizens could draw some small measure of satisfaction knowing that brothers (Nora's brother was one who joined the RAF), fathers and cousins away in the services were 'dishing it out' in good measure. As Air Marshal Sir Arthur Harris, who on 22 February was appointed C-in-C RAF Bomber Command, said, 'The Germans entered this war under the rather childish delusion that they were going to bomb everybody else and nobody was going to bomb them. At Rotterdam, London, Warsaw and half a hundred other places, they put that rather naive theory into operation. They sowed the wind and now they are going to reap the whirlwind…'

In April, German rhetoric was quite different. Walter Darré,[38] German Minister of Agriculture, said: 'As soon as we beat England, we shall make an end of you Englishmen once and for all. Able-bodied men and women will be exported as slaves to the Continent. The old and weak will be exterminated. All men remaining in Britain as slaves will be sterilised; a million or two of the young women of the Nordic type will be segregated in a number of stud farms where, with the assistance of picked German sires, during a period of 10 or 12 years, they will produce annually a series of Nordic infants to be brought up in every way as Germans. These infants will form the future population of Britain ... Thus, in a generation or two, the British will disappear.'

Endnotes Chapter 4

36   In all 69 people died and 89 were injured as 112 HE and machine gun fire reigned down on the city.

37   An excerpt from the book 'Norwich Under Fire' by George Swain states: 'By the end of 1944 we had more than 1,450 alerts. 670 high-explosive bombs and 25.000 incendiary bombs: 330 of us had been killed and 1,100 wounded: 30.000 homes had been damaged, more than 2.000 of them beyond repair.'

38   Walther Darré, born 14 July 1895 at Belgrano, Buenos Aires, Argentina, was arrested in 1945 at Flak-Kaserne Ludwigsburg and tried him at the subsequent Nuremberg Trials as one of 21 defendants in the Ministries Trial 1947-1949). He was charged under seven counts and found guilty on three of these, including atrocities and offences committed against civilian populations between 1938 and 1945. Darré was sentenced to seven years but was released in 1950. He died on 5 September 1953 of liver cancer.

Chapter 5

Jump For It!

To Hell with submarines
They get blown to smithereens
And keep your bayonet chargin'
It gives little safety margin,
Armoured tanks are very fine
'Til some idiot lays a mine
And fighting ships at sea
Would scare the pants off me.
If heroics are the need
The words that one should heed
Are 'Be backward coming forward'!
Be a calculating coward
And with methods of the day
Fight the safest way
And this by implication
I assess as aviation
Since Irvin was so cute
To invent his parachute.

***Thanks to Leslie Irvin* by Jasper Miles**

When 21-year old Bernard John Kemp joined the RAF in May 1940 he 'put in' for an observer's course, although it was his ambition to be a pilot. Above that ambition, however, was an impatient desire to get really flying and into the war. The authorities had informed him: 'You can get an observer's course in six months, but you will have to wait a year to go for pilot.' Kemp, therefore, went through all the normal routine agony of the Initial Training Wings at Hastings and the observer's course at Prestwick. Completing training at Louth in Lincolnshire on the first course to be held there, he remembered with some vividness the remark made by the officer who finally pinned upon him his observer's brevet. Said the officer: 'This is too easy. If I had my way, you chaps would do ten operational trips before you got one of these.' Precisely why any member of an aircrew, except the pilot, should be denied a brevet before having 'gone operational' escaped Kemp. The officer concerned however, need not have worried about young Kemp's justification of his badge in any way whatsoever. He was shortly to have more excitement than is normally crammed into ten operational trips, however arduous. At 21 he was a quick-thinking, highly imaginative man with a keen sense of observation. In the early days of his operational flying on Whitleys on 58 Squadron at Linton-on-Ouse, he noticed that some of the girls around, locally known as 'chop' girls, seemed to have a 'jinx' on aircrews. More than one of them was known to have been the special girlfriend of quite a string of men who went off in their aircraft and never came back. Kemp forced himself to go out with the one who had the most impressive list, just to show that he was not scared and felt as jumpy as the devil about it for some time afterwards.

Meanwhile, he was keen on his job and found astro-navigation absorbingly interesting. Much of the time when the other boys were beating things up at dances or at parties, he would be hanging out of the window of his billet, shooting star positions with his sextant. As a result of this practice, there came a certain evening - 7/8 September 1941 - when he was the only man on 58 Squadron to get his aircraft over its objective (Berlin) on time. The rest of the 196 aircraft (103 Wellingtons, 43 Hampdens, 30 other Whitleys, ten Stirlings, six Halifaxes and four Manchesters) were delayed in reaching the target area owing to inaccurate wind-forecasting.

Kemp was observer on Whitley V Z6947 GE-S flown by Sergeant Kenneth McIntosh Newton which took off from Linton at 20.15 hours. Fortunately, it was a warm, balmy evening because Kemp's lively imagination had been bothering him. When he went to draw his Irvin suit, he noticed that the storekeeper had issued him with a second-hand one on which was marked the name of a man he knew to be dead. He tried to hand it back and asked for another, but the storekeeper inquired what all the bother was about and was there anything in that to make anyone 'windy'? To accept any such suggestion, or admit it even by inference, was more than any 21-year old of self-respect could face. Kemp, therefore, took the dead man's suit without another word and went off. But he didn't take it to the aircraft; he stuffed it in a kit-bag in his quarters and left it there when the time came to take to the air.

At this time Kemp was a seasoned man with seven completed operations behind him. He was not suffering from 'beginners twitch' and, as he went out to the aircraft at its dispersal he was in a normal frame of mind. Kemp was probably one of the most normal types of people you could have found in circumstances of this kind. He was a stable character who did not regard himself as any form of hero and would be both horrified and embarrassed if anybody suggested any such thing. Nevertheless, a lively imagination is not always a pleasant thing to have and once again that evening Kemp found his own inconvenient. As the big Whitley aircraft started rolling and then began to gather speed down the runway, he saw a line of people standing at the grass edge. Undoubtedly they were just the ordinary people who either by duty or normal curiosity are on hand at any time of operational take-off in war. But, as the aircraft gathered speed and flashed past them, Kemp saw that they were all waving and he had a sudden desolate conviction that they were waving the aircraft and its crew a final farewell.

Those who like to deal in omens may find food for thought in this. Kemp admitted that he never remembered having had that impression on any previous take-off. He excused him by saying that at the time he was 'a bit keyed up' or the point at which the stable man finds added strength and the unstable has hysterics.

From that point however, Kemp was far too occupied with the business in hand to let omens and superstitions trouble him. The Whitley had as heavy a bomb load as was possible for its long trip to Berlin and back. It was carrying two 1,000lb bombs, four 500-pounders and a rack of sixty incendiaries weighing four pounds each. By later standards of the war, such a load was a pip-squeak but the Whitley had been designed long before the war when targets at such a range as Berlin were not thought of seriously as objectives. Meanwhile, Kemp concentrated on his navigation.

The Whitley arrived over the German capital dead on briefing time - 23.55. It was the only aircraft on 58 Squadron to get there with such perfect punctuality. But directly it arrived, the welcome it received from the ground gunners was more than enthusiastic. As Newton steadied the aircraft to make the run-up to his target, Kemp was conscious of a tremendous noise of bursting flak all around; constant flashing explosions and clouds of smoke wreathing back in the strong moonlight. The reception was as hot as he had yet known in his experience and he privately reckoned that they were going to be lucky if they got away without serious damage after they had spotted their precise target and dropped their load. Then, suddenly, he reared violently in his seat and hardly knew what had happened when he found he had collapsed back again. The first sensation had been as though some unseen giant had given him a terrific kick in the backside, but apart from the stunning impact which knocked his senses askew, he felt no pain. He realized that he was bleeding and a gingerly investigation showed him that he had been wounded in the right buttock. This was unpleasant, but as he began mentally to 'get back' to his surroundings, he realized something that was positively uncanny. All the noise, which had previously been so deafening, had now stopped. There was an eerie silence. It was suddenly borne in upon him that even the engines had stopped their strong, heavy drone. For a moment he had the unreal impression that the aircraft was empty. Then he saw that Newton and co-pilot, Pilot Officer F. R. Wilbraham, were still in their seats. At the same time he realized that the plug which connected his helmet ear-hones with the intercom was dangling clear. He saw that, in the wild jerk he gave after being wounded, he must have dragged the plug out of its socket. And as a result he had not heard his pilot's order, which had obviously been given, for the crew to bail out.

As clear thinking came back to him, Kemp decided that since the disabled and impotent aircraft was plumb over the centre of the city of Berlin, it was time for him to see that the object of the flight was carried out. Therefore, he released the bombs. As the aircraft lifted to the loss of their weight, he saw Wilbraham throwing him down his parachute pack. After which the 'second dickie' took a firm hold on his own release ring and dived away. Kemp dragged himself close to the captain's seat and yelled out to know if there was anything he could do. Newton had his feet braced up in the 'crash' position. He shouted back that there was nothing anyone could do and told Kemp to get away as quickly as he could. Kemp dropped down to sit in the edge of the escape hatch with his feet dangling into space while he tried to tighten his straps. He took a dive, but jerked at his rip-cord almost too soon. As he turned head over heels in space he had the alarming sight of his pilot-chute narrowly missing the tailplane of the aircraft as it whisked overheard. A few minutes later Newton was dead, killed in the final crash when the Whitley struck the ground and disintegrated in flames. [39]

Meanwhile, for a few moments, Kemp went through a spasm of rather sickening pain. This was because the parachute harness around his body was looser than it should have been; the result of wearing no Irvin suit, because he would not wear one that had belonged to a dead man. The webs of the harness which passed down between his thighs therefore caught him awkwardly when the speed of his free fall was checked by the opening envelope above. The resulting pain was just as bad as many a rugger player or cricketer has had cause to remember. When a man's weight is almost entirely depending upon thigh straps of this kind, it isn't too easy to alter position and gain relief. Kemp heaved and hauled himself on his shoulder straps as best he could, through sickening moments until he found some sort of ease. He realized that he was as brightly lit up as any music-hall star on the stage of the palladium. First, one searchlight found him. Then, as it steadied on the white mushroom of his parachute, others came swinging in vast sweeps across the sky to join the focus of the first. The pyramid of searchlights gradually followed Kemp down to the ground. He had a feeling of helpless nakedness in that glare as he came down the last few thousand feet. The AA guns were now thudding and banging a ferocious chorus all over the city as the rest of the raiders began dropping their bomb loads. As the vague darkness of the ground approached, he tensed himself, waiting for the sudden barking of machine guns, which he more than half expected would open up and riddle him with bullets when he came within range.

However, for once the German gunners restrained their well-known sense of humour. Not a shot was fired at him as he dropped down belie the trajectory of the searchlight beams and fell quickly into what appeared like a mass of sea anemones, an area of thin, waving fronds which seemed to reach out to grasp him.

Kemp used his common sense and decided that it must be a forest. But it wasn't. As an example of the extraordinary tricks that the eyes of a sorely-strained man can play, it was just ordinary medium-length grass, with very hard, solid earth directly beneath it. Kemp hit the earth with a force that knocked the breath out of him, long before he reckoned he was within 20 feet of it/ the thud caused him to bit his lip, heavily and painfully. But he still kept his head, plunged and struggled to free himself of his parachute harness before dropping down, relaxed, to regain his breath.

He was very quickly surrounded by a group of men who had evidently watched his descent in the searchlight display and were ready at his point of landing. From what he could see of their dress, they were farm workers. They treated him decently enough, helping him down into the cellar of a house nearby, where they were soon joined by some of the women of the establishment.

The place where Kemp had landed was just east of Berlin, in agricultural country. He spoke very little German, but found it easy enough to make the farm folk understand that he was wounded. One of the men glanced at his blood-stained seat and gestured to him to take his trousers down but this he refused to do in front of the women. Somehow, there was a difference between making a public show of oneself and being treated by a trained hospital nurse. Both the men and the women seemed amused at his modesty, but they helped him to another cellar where a couple of the men undressed him. Then one of the women appeared with hot water and rough bandages and did her best for him while he was lying on his face. This woman earned his undying gratitude, for, during the four hours that he stayed there and before men of the Luftwaffe arrived to take him away, she smuggled some chocolate to him. The men treated him decently too, to his surprise. They all gave him cigarettes and when he indicated he was hungry, brought him black bread, a coarse jam and some ersatz coffee, which was peculiarly bitter to his tongue. Apparently there was no sugar or milk available.

At last, a very young and smart officer of the Luftwaffe put in an appearance and saluted him with considerable ceremony. This man spoke fair English and remained just as polite and pleasant when Kemp announced that he was a sergeant and not the holder of a commission.

Throughout all this the shrewd Kemp suspected the usual build-up in which the confidence of captured men was very carefully gained by pleasant treatment in order that they might be persuaded to talk. He was not far wrong, for after humane treatment of his wounds he was taken to Dulag Luft near Frankfurt-on-Main, where, after refusing to give more than his name and number according to regulations, he was forthwith clapped into solitary confinement. Here he got very thirsty and asked for water. The guards brought him a rough-looking wine, but having heard about such things, he refused to drink it, knowing that it probably contained a relaxing drug that would put him off his guard. Later on he discovered that others were given the same treatment but made the mistake of drinking the wine and to discussing the type of aircraft in which they had been flying. It is a measure of Kemp's character that, even after three and a half years in Stalag Luft I at Barth near the Baltic coast in Pomerania, he was still periodically up for questioning and never disclosed any further details.

Of his life in prison camp throughout the years that followed, Kemp had little to say, other than a general description given by most men who have endured it. 'Oh well - it was hell, but somehow we got through.'

Sergeant Kemp got down on to British soil at Ford in Hampshire at 3 o'clock in the afternoon of 14 May 1945 and walked into his home in the district of Addington, Surrey at 3 o'clock in the afternoon two days later. He became a civil servant, happily married and the father of two young daughters. [40]

Endnotes Chapter 5

39   The two Whitleys that were lost this night were both claimed north of Berlin by Leutnant Rudolf Altendorf of 2./NJG3. Thirteen other aircraft - 8 Wellingtons, two Hampdens, two Stirlings and a Manchester also FTR. Good bombing was claimed by 137 crews in clear visibility. Berlin reported most bombs in the Lichtenberg and Pankow districts, which are east and north of the centre. Damage was reported to several war-industry factories, transport and public utilities and public buildings as well as a zoo, 16 farms and 200 houses. 36 people were killed, 212 injured and 2,873 people were bombed out, some only temporarily. *The Bomber Command War Diaries; An Operational reference book 1939-1945* by Martin Middlebrook and Chris Everitt (Midland Publishing Ltd 1985, 1990, 1995).
40   Adapted from *Jump for it! Stories of the Caterpillar Club* by Gerald Bowman (Evans Brothers London 1955). Gerald Bowman was a novelist who wrote *Sawdust Angel, Pattern In Poison Ivy* and *The Quick and the Wed*

Chapter 6

The Night The Fuel Ran Out

'...Berlin. This attack was not successful owing to 10/10ths cloud over the area. Quite a number bombed on the estimated position; others bombed alternative targets and a number joined in the rover's party around the Ruhr on the way back... one hundred and twenty seven aircraft took-off on this mainly unsuccessful expedition and twelve are missing. About 200 tons of bombs were dropped somewhere in Germany or Occupied France, however.'

The 3 Group Officer who wrote up this plain statement of fact with a cynical overtone for the night of 7/8 November 1941 makes no bones about that fact that the Berlin operation proved a failure. The Command Battle Order included 169 aircraft whose crews were briefed to attack Berlin: 1 Group 22 Wellingtons; 3 Group commanded by Air Vice-Marshal Sir John Baldwin, highly respected and with a reputation for speaking out, 69 Wellingtons and seventeen Stirlings; 4 Group 42 Whitleys 10 Wellingtons 9 Halifaxes. This was the biggest force sent to Berlin up to that time and a further 200 Hampdens, Manchesters, Halifaxes and Wellingtons were dispatched to Cologne, Essen, Mannheim and smaller targets including the Channel Ports. While weather forecasts spoke of cloud over the Ruhr the more severe weather lay to the north and especially in the area of the Bight and northern Germany generally. For the return conditions were expected to be good with clear visibility at the bases.

Pocklington airfield in 4 Group in Yorkshire was set in a small farming community; it had three pubs, an old Anglican church and two rather ordinary hotels, which would serve customers a very restricted wartime meal for five shillings. This was the top price allowed by the Government to any restaurant in wartime. The station itself was not a comfortable billet and in the bleak days of early November 1941 men shivered in the hutted accommodation common to many wartime airfields. 'It was a dreary camp' wrote Squadron Leader John Searby, commanding 405 'Vancouver' Squadron, which was equipped with Wellington IIs.[41] 'The Officers Mess was short on warmth and comfort: I slept in a wooden hut where the draughts from the cracks between floor and wall caused my jacket, which hung on a peg, to swing gently to and fro until I plugged the gaps with newspaper. Accustomed to a dry cold the bitter winds of that first week in November tried the Canadians sorely and relief was found in bottle form - the party spirit was well to the fore and everyone joined in. A long way from home and stuck in a situation over which they had no control they accepted the weather, the wooden huts and the war with a cheerfulness which did them credit!'

On 39 nights between 30 July and 31 December 1941 RAF bombers operated in weather variously described as 'bad,' 'extremely bad,' 'very poor visibility,' 'thick cloud,' 'icing and ten-tenths cloud.' It was cloud that prevented accurate bombing by 82 aircraft that were dispatched to Hamburg on the night of 31 October/1 November when Squadron Leader John Searby flew with Sergeant Edwin John Williams RNZAF on his 'second dickie' trip.

'An attractive personality, he was the very stuff of which bomber captains were made - determined to force his way through regardless of weather and enemy opposition to drop the 3,000lb load of high explosive on the docks which lined the River Elbe. The weather deteriorated as we neared Heligoland and we altered course for a point to the south-east of Hamburg prior to making our final approach to the target. In thick cloud and heavy rain we sought blindly for a reference point, dropping to 4,000 feet over the German coast and encountering a curtain of light flak fire from the defences. The stuff whizzed past the Wellington - red and green blobs of incendiary shells, close and nasty, any one of which might have done for us all, but Williams was going to have his pinpoint and eventually [Pilot Officer Wilson] the navigator, lying prone in the nose, declared he was satisfied and up went the nose in a steady climb back to bombing height. At 10,000 feet we were in thick cloud again - the captain flying very accurately on his instruments as if engaged on a training exercise.

'We must have flown into a huge cu-nim because the next few minutes brought nightmarish conditions with lightning flashes, torrential rain and upcurrents of great strength which tossed the Wellington around the sky in the most sick making fashion but Sergeant Williams stuck doggedly to his instrument panel, reassuring the crew from time to time and expressing the belief we should come out of it very soon. We did so, to be grasped by a forest of white beams and engaged by heavy flak accurate for height and close at times - too close for my peace of mind and all of which, according to our stout captain, confirmed our nearness to the target. More heavy cloud and the navigator piped up to say we were only a matter of two or three minutes away from the dropping point and he was ready to go back into the nose and do the business (no bomb aimer was carried at this time and the navigator did everything). To my dismay, Sergeant Williams cut into the navigator's announcement and said he would not drop on this run unless they had the docks in the bombsight: he would do another run losing a couple of thousand feet in the hope of seeing the ground! Silence in the aircraft - from the crew that is - and a sheet of cloud hid us from the searchlights - a blessed relief. Then he was addressing me - would I kindly go to the rear of the aircraft and operate the hand pump which replenished the oil tanks in the engine nacelles?

'Just a few strokes, Sir, we have to do this from time to time (apologetically).'

'Of what kind of metal was this youngster made?

'At 8,000 feet we began a second run towards the area where we expected the docks to lie and the ground gunners redoubled their efforts; both heavy and light flak streamed up from below and again we drove into a cloud layer with a few gaps.

'Can you see the docks Freddie?' [Flight Sergeant Joseph Raymond Frederic Bourgeau RCAF] from the Captain.

'I can see the river - can't we get the hell out of this... OK keep her straight - we're heading directly towards them... left a bit ...steady, steady ... bombs gone.'

'And that part of the business was over. I felt the vibration from beneath as the thousand pounder and the four 500lb bombs left the racks. Then we were in thick cloud and the nose was pulled up sharply as the shells burst around us... up, up until the Wimpy was standing on its tail... the stick went forward and Williams put her in a steep dive turning out toward the coast but holding the nose down until she commenced shaking violently and I wondered if the airframe would hold out much longer. We lost the flak for a brief spell but they picked us up once more as we neared the estuary of the Elbe where Williams put up another acrobatic feat - then silence as we flew out over the North Sea. What a display of guts from that young New Zealander, was the thought in my mind. There was no other aircraft in the vicinity - we had the whole of the Hamburg defences to ourselves to the best of my belief and it was a very dicey affair from first to last. We landed back at Pocklington and I thanked the young captain for the ride.

'Sorry about the rough stuff,' he grinned, but these Wimpys can take it - I'd rather fly one of these old crates than anything else.' A few days later, on 7 November, Berlin was laid on.'

The noon weather forecast on 7 November showed that there would be a large area of bad weather with storms, thick cloud, icing and hail over the North Sea routes by which, the bombers that night would need to fly to Berlin, Cologne and Mannheim and back.[42] Even so, the aircraft were allowed to take off. It was an important milestone for Bomber Command, the C-in-C, Air Marshal Sir Richard Peirse KCB DSO AFC, having ordered a major effort with Berlin the main target. On the morning before the raid, the various bomber group planners got together over the conference telephones to consider the forthcoming night's work. First they heard what the met officers had to say. The forecast was by no means optimistic. Clouds would be building up to 15,000 feet over Germany and icing conditions were prevalent. A strong wind would be helping the bombers towards their targets, hindering then on the way back. It was not a good forecast, but it was typical of many nights during that appallingly bad winter of 1941/42. It did little to allay the fears of many, Squadron Leader John Searby among them.

'As I sat in the briefing room at Pocklington I listened to the Intelligence Officer armed with the latest information on the gun and searchlight zones. It was all matter of fact and nobody batted an eyelid. The ten crews of 405 Squadron sat passively accepting a repeat of that which they already knew. On the big wall map of Europe the green tapes crossing the North Sea to Denmark and on to eastern Germany, where the huge splodge of hatched green and red areas surrounded the target, drew their attention - and it seemed a very long way to travel. There were no questions - it was old stuff, anyway - and we passed on to the Met briefing. Outside it was a bitterly cold day with the sun occasionally breaking through yet there was no hint of bad weather so that when the Meteorological Officer hung up his chart showing a cross section of what was expected I received a jolt - just one thing caught my eye - the pillar-like cu-nim clouds standing in regular formation over the sea route to Jutland.

'Forecast temperatures were low and there was evidence of frontal activity off the Danish Coast. All in all I thought it likely to be a rough night. The convection clouds were not expected to rise much above ten thousand feet and there was risk of ice in such clouds. Over Germany heavy cloud could be expected with occasional breaks: this cloud could extend to ten thousand feet. Bomb loads for the Wellington IIs of 405 Squadron were 1 x 1,000, 4 x 500, 1 x 250 and two small bomb containers each holding ninety of the little incendiary bombs: no bad effort considering the distance to the target. The route to the target lay directly east from the Yorkshire coast to Denmark - for all No 4 Group aircraft whilst Nos. 1 and 3 Groups flew almost directly east from the Wash and Norfolk coast to the 'Big City' following what was to become a well-worn track skirting the major cities of Osnabrück, Hannover and Magdeburg.

'Only the previous night I had attended a party in the ante room to welcome two visiting members of the Canadian Parliament - an occasion for everyone to speak up without fear or favour - and heard not a single voice raised in complaint... A different mood prevailed at our pre-flight supper and there was no enthusiasm for the task ahead. The weather map had done the trick; those tall 'thunderheads' over the North Sea spoke a language well understood by all save the newest arrivals - and the latter were not represented in the tally of ten crews headed by Flight Lieutenant John Fauquier in command of 'B' Flight. He was an officer and commander of distinction with a reputation for bluntness'. Fauquier had been a bush pilot in the 1930s, a splendid flyer who had flown in all sorts of aeroplanes in the northern territories. He was really experienced and was regarded as a real warrior. He had made a name for himself as a tough individual of few words; his curt manner meant that he did not have to say a great deal. Fauquier wrote: 'At briefing we were told our target was to be Berlin. When the Met Man came along with his chart I immediately smelled a rat: he was nervous and seemed unable to make up his mind about the wind velocity for the return to base.' [43]

The risks in reaching Berlin, involving a round trip of 1,200 miles for British bombers, were great at any time and many of the aircraft that night, with a full bomb-load, would have very little range to spare. Aircraft in general use at the time were the 'old guard' of Bomber Command: the slow, tough and very reliable Whitley, the equally dependable Wellington and the Hampden. There were only a handful of the newer Halifaxes, Stirlings and Manchesters. The planning conference that morning must have considered all this and particularly the fact that the limited range of the Whitley made Berlin a critical target in the best of weather. Even on good nights the bombers normally based in northern England often took off from southern airfields to save fuel. The planners must also have discussed the probability that these aircraft would use up extra fuel by climbing to clear the forecasted cloud-tops. There was little chance of flying through the clouds because of the severe icing forecast. One who recalls the events of this dramatic day is AVM (later Marshal of the RAF Sir John) Slessor, AOC of 5 Group. 'I had been away visiting Bomber Command HQ that afternoon and on returning to my headquarters was met by our senior meteorological officer, Mr. Mathews, who told me in no unmeasured terms that in his view we should be undertaking a quite unjustifiable risk in sending the Hampdens to Berlin in the icing conditions to be expected. I had the utmost faith in him - he knew the aircraft and the crews and what could be expected of them almost as well as I did and I had no hesitation in diverting my group from Berlin to a closer alternative target and so informed Command.' Slessor telephoned the C-in-C that afternoon and stated that his Hampdens lacked the range to fly to Berlin and back given the forecasted strong westerly winds. No doubt frustrated by the recent long run of bad weather and poor bombing results, Peirse wanted to mount a major effort against Berlin but he allowed Slessor to withdraw his 61 Hampdens and 14 Manchesters from the Berlin force and send them to Cologne instead. The target for 53 Wellingtons and two Stirlings of 1 and 3 Groups was Mannheim, which left 169 aircraft of 1, 3 and 4 Groups: a hundred and one Wellingtons, 42 Whitleys, 17 Stirlings and nine Halifaxes, to raid Berlin.

Labouring under a maximum load of bombs and fuel they took off from airfields all over Eastern England and climbed painfully up through the gloom. Many had been briefed to land at southern airfields as there was little chance of them getting back to their own home bases. The first difficulties came soon. The weather had worsened in one slight respect; there were some cloud tops which were higher than the 'solid top' forecast. The slender margin by which the bombers would reach Berlin was cut and the heavily loaded aircraft were faced with a problem. They could fly through the clouds and risk icing up, which with their heavy load might prove critical, or they could fly round the cloud tops and use up their precious fuel.

Flight Lieutenant John Fauquier was one of many who carried vivid memories of that night. He wrote: 'Take-off was around 11 o'clock that night. All went well on the way out until we climbed out of the overcast which was solid all the rest of the way so that we had nothing but dead reckoning and forecast winds to get us to the target. Finally, we reached the point where we thought and hoped Berlin lay beneath - dropped our bombs and turned for home. It wasn't long before I realised we were in trouble because the winds had increased greatly in strength and were almost dead ahead. Eventually, I lost height down to a few hundred feet - to avoid icing conditions and to save fuel since the head wind would be less strong.'

Pilot Officer Harry Drake, making his second operational trip as second pilot in a 10 Squadron Whitley also carried equally vivid memories of that night. Drake, who later became an Air Correspondent in London, recalled that on the outward trip they were able to get their Whitley about 1,000 feet above the solid cloud. But stretching at least 4,000 feet higher were other cloud tops. 'We knew icing to be heavy and decided to go round the clouds. But we lost so much time and fuel in doing this that, by the time we had crossed into Denmark and the solid cloud had begun to break up, we knew that it would be useless trying to reach Berlin as we hadn't enough fuel left. Instead we searched along the north German coast and found Lübeck sufficiently clear for a visual bomb run and we bombed there.'

What happened to Drake happened to many others and aircraft which should have gone to Berlin diverted to Lübeck, Warnemünde, Rostock. Kiel, Schleswig, Sylt and other towns because they knew they would have insufficient fuel to reach the German capital. But their difficulties were not over. 'We were still worried about the development of upper cloud,' says Drake of the return journey 'and of the difficulties of making a descent through cloud when we got back home. Conditions then were not what they are today and it would have been more than tricky making a straight down blind approach with hills round the airfield! So we looked for a 'hole' off the Danish coast and luckily found one. We spiralled down through it and found the bottom of the cloud only 500 feet above the sea. We returned all the way at 200-300 feet, met frequent thunderstorms and grand displays of St. Elmo's fire, although we were in no position to appreciate it! Most of our crew were sick all the way back.'

But Drake and most of the other's who had diverted from Berlin, had the fuel reserves left to make a British airfield. For those who struggled through to Berlin it was a different matter. In the first place, most of them flew straight through the cloud tops on the outward journey; they had to, otherwise they would not have had sufficient fuel left for the return journey. In the high cloud tops many iced up. Some got into difficulties and by the time they should have flown out of the cloud tops they were below the level of the solid cloud.

For those who reached Berlin another surprise was waiting. A clear area had been forecast for the target region, but Berlin was far from clear. Not that this had any bearing on the ability of the aircraft to get back. Its main effect was to make bombing difficult. It was on the return journey that troubles began in earnest. One of the men who got to Berlin and back was Sergeant D. K. Brearey who after the war ran his own business in Ipswich. Brearey made the trip in the rear-turret of a Whitley on 102 Squadron. He recalled that there was light flak over the target but no sign of enemy fighters. After a fairly successful bombing run Brearey's aircraft ran into bad weather. It was blown off course and lost its way and drifted over Hamburg where it was met by a lively flak barrage.

'Near the Dutch coast,' recalls Brearey, 'we ran into dense cu-nim cloud, iced up and both our engines cut out. My captain, Sergeant Tony Wickham, took the Whitley down in a rapid glide from 12,000 to 2,000 feet. In the centre of the cloud the wireless operator got a shock from his set, great chunks of ice flew off the props and wings and hit the fuselage. My turret was completely frosted over and I could see nothing. It suddenly became very cold and very still.'

The Whitley levelled off at 2,000 feet and Wickham managed to nurse the engines back into life. It was bad weather all the way back to base, but the Whitley made it safely. Others in the Squadron were, according to Brearey, so tired after the trip that they crashed on landing. One ploughed through a group of Nissen huts. 'These casualties were in addition to those officially listed as missing,' says Brearey.

From all over Bomber Command came stories of the hectic night of 7 November. Crew members were unable to touch metal parts without their gloves. A navigator who laid down his gloves for a few moments was unable to put them on again as they had become frozen solid. Sixty-six degrees of frost was recorded inside cabins and icicles grew around chins like brittle white beards. But the majority of aircraft managed to overcome these hazards in one way or another. They used their superchargers to climb quickly above the icing level, they spiralled down to sea-level or they tried to fly around the worst areas. And all these manoeuvres used up valuable petrol. [44]

At Pocklington John Searby observed that the Canadian crews on 405 Squadron returned 'utterly fatigued, half frozen and disgusted.' 'Winston Churchill's succinct phrase about not fighting the weather as well as the enemy' came to mind. It was a view shared by Flight Lieutenant John Fauquier whose Wimpy was heavily damaged by flak but succeeded in returning to a crash landing.

'I have seen the North Sea in many moods but never more ferocious than that night. Huge waves of solid green water were lifted from the surface and carried hundreds of feet by the wind. After what seemed like hours in these appalling conditions I realised we were unlikely to make base. I had little or no fuel left and told the crew to take up ditching positions though our chances of pulling off a successful ditching in darkness amid that hell of strong winds and blown spray were practically nil. It was then I saw briefly one of those wonderful homing lights and made a bee-line straight for it, crossing the English Coast with all gauges knocking on zero. In a few moments we found what I thought was Driffield but it proved a non-operational airfield. I could just make out the runways - as it was early dawn - and slapped the wheels and flaps down whilst I still had power, only to find at a hundred feet the runway blocked with railroad iron to prevent the enemy making use of it. They had erected pylons of this stuff all down the runway but, of course, they forgot to put any kind of obstruction on the grass surfaces. I landed but swerved to port and damaged the starboard tail plane. All in all we were lucky and nobody was hurt, thank God. As soon as we climbed out of the aircraft we were surrounded by the Home Guard, who were most hostile in spite of our uniforms and the RAF roundels on the fuselage. They were going to lock us up! I asked repeatedly for the Colonel, or whoever was in charge, to make contact with Pocklington but he was still suspicious. He asked me when we had left England and when I said, 'eleven o'clock last night,' he replied: 'no aircraft can stay in the air that long.' Finally, everything was smoothed over and we were picked up and returned to Pocklington.' [45]

'Fauquier's luck' recalled John Searby 'held on this and other occasions but he was still a bit sour when I met him in the Mess the night following the Berlin raid, expressing himself forcefully on the subject of Air Staff planning at Headquarters Bomber Command. Although he makes no mention of it his Wellington was heavily damaged by flak: the misery of that North Sea crossing with a 70 mph headwind eating up the scant fuel reserve and the expectation of a ditching in darkness and a wild sea, was shared by all save the Halifax crews who had no fuel worries and whose new four engined bombers carried them over the weather. What happened to the other nine Wellingtons? The weather encountered is summed up briefly as 'Bad; 10/10 cloud with few breaks' yet six crews battled their way to the target area and two caught a momentary glimpse of the city - 'Q' captained by Sergeant Sutherland the 'G' captained by Sergeant Suggitt: whilst others bombed on the estimated position. Flight Sergeant Alexander Lawrence Dennis Hassan RCAF, flying Wellington 'D' dispatched a signal to base at 0223 'Operation completed' but of this gallant crew nothing more was heard and they may well have fallen victims to the fury of the North Sea, as did others when the petrol ran out. Sergeant A. N. McLennan's Wimpy was hit in the front turret, tailplane and starboard engine by flak but the Canadian pilot made it home.

Sergeant Edwin Williams abandoned his sortie with zero oil pressure on one engine 'but typically, jettisoned his bomb load over what he believed to be Wilhelmshaven before staggering back to base on one engine.' The New Zealander estimated the wind at height to be 80 mph with an outside temperature of -42C. Their slow crawl back to base with the Wellington capable of not much over a hundred and ten with one engine feathered and tending to overheat called for skill and determination but as he put it 'it was nothing to write home about - the alternative was a ducking.' Williams, who Searby had flown with on his 'second dickie' trip a week earlier, jettisoned his bomb load over what he believed to be Wilhelmshaven before staggering back to base on one engine.'[46] 'Determined to force his way through regardless of weather and enemy opposition' Williams was killed just over a month later, on the night of 28/29 December when his Wellington was shot down on the trip to Emden.[47]

Halifax crews on 76 Squadron at Middleton St. George, which dispatched five aircraft, described the weather thus: 'as a result of the extreme cold Air Speed Indicators and Compasses were made unserviceable'. The alcohol in the compass bowl was adjusted to withstand very low temperatures indeed. The powerful four-engined Halifax Is, not long in Squadron service, rode above the threatening cu-nims but out of the six aircraft detailed only two, Pilot Officer Calder and Sergeant Herbert reached the Berlin area and bombed from 15,000 feet. Squadron Leader Hillary and Sergeant E. T. Borsberry both abandoned their sorties, the former because his port outer engine ran continuously on rich mixture and the latter because his instruments were frozen up. Squadron Leader Bouwens abandoned his primary target and bombed Flensburg. Calder and Herbert flew for eight hours that night whilst the humble Whitleys struggled on for eleven hours before reaching their home airfields. Sergeant Lloyd-Jones and his crew on 78 Squadron at Middleton-St-George were fortunate in lobbing down at RAF Coltishall in Norfolk with only two gallons remaining in the petrol tanks after throwing various items into the sea. This included the four Brownings from the rear turret 'owing to a misunderstanding by the rear gunner'. Nevertheless, the action saved the lives of Sergeants Connolly, Wilson, Beaton and Cragg plus the captain. Sergeants John William Bell and South African pilot Ernest Johnston Sargent with their crews, on 78 Squadron, did not make it - the last message from the first named stated 'task abandoned at 0225' which was the time they would be over Berlin or near it but could see nothing owing to the dense cloud. Bell and his crew were killed; Sargent also died but his crew were taken prisoner. Pilot Officer Charles Acton Chaplin Havelock on 77 Squadron at Leeming reported 'severe icing from 2-10,000 feet' but he made it home. Havelock and all except one of his crew were killed on 27/28 December on the operation on Düsseldorf.

At Dishforth 51 Squadron lost two Whitleys. Z9130 flown by Squadron Leader Peter George Scott Dickenson was lost and only one member of the crew survived to be taken prisoner. Sergeant A. W. MacMurray RCAF and crew on Z6839 crashed near Bierum in Holland and were taken prisoner. At Topcliffe 102 Squadron lost three Whitleys. On Z6796 captained by Sergeant Reginald Charles Matthews his wireless operator called for a QDM ('What is my magnetic course to steer to base airfield?) at 0726 hours when they must have been only a short distance from the Yorkshire coast since other aircraft on his squadron landed at 0700 and 0743 respectively. Matthews' and his crew were lost without trace. The two other Whitleys on 102 Squadron that were lost were flown by Pilot Officer Bruce Buchanan Percival Roy and Sergeant Thomas Henry Thorley. All ten crew were lost without trace. One of these aircraft was 'fixed' by W/T as being sixty miles north of the island of Borkum at 0642 that morning but never reached base. This position and time put him on track for England with another hour and a half to fly before crossing the coast but nothing further was received and he too, was probably short of fuel. A similar fate befell Pilot Officer Kenneth Monckton Tuckfield and Pilot Officer Douglas Edwin William Brown on 58 Squadron at Linton-on-Ouse. Tuckfield's last known position put him sixty miles due east of Scarborough heading directly for base, whilst in the case of Brown the Operations Record Book contains the entry 'believed landed in the sea'.

No 1 Group contributed twenty-two Wellingtons for the Berlin attack and a further 60 were detailed to attack Mannheim and the Channel ports. Only nine crews on the Berlin operation claimed to have bombed anywhere near the 'Big City' and four failed to return. These all came from the four Polish squadrons operating under Air Vice Marshal Oxland's command. Three more Wimpys were lost over Mannheim. At Binbrook only Flight Lieutenant D. T. Saville and Pilot Officer P. J. Oleinek on 12 Squadron succeeded in getting to the target area out of the five dispatched. Oleinek was engaged by heavy flak fired through 10/10th cloud from which evidence he deduced that they were in the vicinity of Berlin since it coincided with the navigation plot. He then flew around the area at 12,000 feet but all to no avail and he contented himself with dropping his bombs at intervals wherever the flak was most concentrated before turning for home. This Wimpy flew for eight and half hours before landing safely back at an airfield in Lincolnshire. Pilot Officer A. Heyworth, captain of one of the Wellington IIs carrying eight 500lb bombs to Berlin flew in cloud with tops at 19,000 feet; he experienced bad icing troubles and extreme cold and was forced to return short of fuel. Pilot Officer C. A. Barnes with a battle-tried crew found his guns frozen up and ice so heavy on his Wellington that to continue was impossible, while Sergeant C. Voller experienced oxygen failure.

In 3 Group, 7 Squadron at Oakington dispatched eleven Stirlings for two aircraft missing. American Flying Officer Douglas Byrd Van Buskirk RCAF crashed near Duisburg with the loss of all the crew who were buried at Reichswald Forest. Sergeant J. W. C. Morris RAAF crashed near Hekelingen. All the crew were interned at Crooswijk. The remaining aircraft captains attacked alterative targets or abandoned due to icing or mechanical failure. At Wyton four Stirlings on 15 Squadron reported bombing Berlin on ETA and one bombed Münster. At Feltwell 75 Squadron RNZAF dispatched eight Wellingtons. Sergeant John William Black was last heard transmitting on the Group frequency at 2236 hours - 'target not attacked'. The New Zealander's Wellington crashed at Oldebloom with the loss of all six crew. Pilot Officer W. R. Methven's wireless operator transmitted an SOS at 2228 hours before the Wellington went down. Methven and four of his crew were taken prisoner. Sergeant John Cuthbert McKechnie Gibson RNZAF the navigator was killed and later, was interned at Reichswald Forest. At Honington of nine Wellingtons on 9 Squadron that were dispatched, five reached the target area but none claimed to have attacked the primary due to cloud and severe icing conditions. At Waterbeach 99 Squadron dispatched eight Wellingtons, none of which succeeded in identifying Berlin owing to thick cloud and severe icing. Three aircraft, piloted by Flight Lieutenant J. P. Dickinson, who was killed, Pilot Officer W D Moore, and Pilot Officer Charles George Gilmore, which crashed at Beilen in Holland with the loss of all the crew, all failed to return. On 101 Squadron at Oakington, three Wellingtons attacked Berlin, one abandoned the sortie and the Wimpy piloted by Pilot Officer William Dave Clayton Hardie failed to return. He and his crew were lost without trace. At Marham five Wellingtons on 115 Squadron attacked through thick cloud and two more unable to reach Berlin attacked alternative targets. Its parent squadron on the station, 218 Squadron, dispatched 13 Wellingtons, all of which went the distance though none were successful identifying the target and there were no early returns. At Mildenhall one Wimpy on 149 Squadron attacked Berlin and three abandoned the operation owing to severe icing. Sergeant Stanley William Dane and his crew failed to return. Only one member of the crew survived. At Stradishall 214 Squadron detailed five Wellingtons for the raid. One bombed on ETA, two returned early due to engine failure and one was reported 'missing'. The fifth Wimpy failed to find Berlin owing to cloud. At East Wretham 311 (Czech) Squadron dispatched ten Wellingtons, seven of which claimed to have attacked Berlin. The other three abandoned the operation owing to weather and engine failure.

Only 73 aircraft in total reached the general area of the German capital, where bombing was scattered. The Spandau Power Station was hit and damage was caused to a considerable area of Moabit, one of the working-class quarters of Berlin. In all, twenty-one Wellingtons and Whitleys and Stirlings failed to return and there seems no doubt that most if not all of these were forced down through lack of fuel. Losses on the Berlin raid were high at 12.4 per cent (21 aircraft). Total losses for the 392 sorties flown were 37 aircraft (9.4

per cent); more than double the previous highest for a single night.

It was between six and seven o'clock next morning that the extent of the night's losses became apparent. Most of the attacking aircraft were by this time nearing the North Sea. SOS's began to come into the bomber bases. 'Running short of petrol. Ditching,' was the gist of these messages as, one by one, pilots realised that they would never have the fuel to get back to England. Understandably the most critically placed were the aircraft which got to Berlin. Forty-three aircraft claimed to have bombed Mannheim as ordered. Seven Wellingtons including two Wimpys on 300 (Masovian) Squadron failed to return from this raid. Few bombs fell on Cologne also but all 61 Hampdens and 14 Manchesters returned safely.

One brilliant piece of airmanship and bravery was denied the full credit it deserved because the aircraft ran out of petrol. Wellington X3206 on 214 Squadron flown by Pilot Officer Lucian Ercolani had reached Berlin dropped its high explosive load successfully through a gap in the clouds but before the crew could release their incendiaries the gap closed. The son of an Italian furniture designer and manufacturer[48] who had come to England in 1910, Lucian Brett Ercolani was born at High Wycombe on 9 August 1917 and educated at Oundle, where he excelled at sport. He left school in 1934 to work at his father's company, Ercol. When war broke out he joined the RAF and trained as a pilot in Canada, returning in May 1941 to join 214 Squadron. Knowing that they would fly over other worthwhile targets on the way back, Ercolani kept the incendiaries, unfortunately, as it happened. Over Münster a lucky flak burst hit them and set the incendiaries alight. The bomb release gear was smashed and there was no getting rid of the burning load. The wireless operator took ten minutes to fight his way through the smoke and flames to reach the rear of the Wellington, whence he could let out his trailing aerial. This would give him a greater range and increase the chances of maintaining contact with base. Before long the Wellington was ablaze along the whole length of its bomb bay. The crew used fire extinguishers and coffee from thermos flasks to try to put the fire out, but to no avail. The intercom had failed and Sergeant Fry the rear gunner was effectively cut off from the rest of the crew by the flames but the aircraft still flew. After a while the fire subsided but the incendiaries still burned in the bomb-bay. Realising that it must have been a great help to the AA gunners who continued to shoot, Ercolani closed the bomb-doors to hide the tell-tale light. Still with the fire in its belly, the Wellington crossed the coast and started out over the sea. It lost height steadily but there seemed a hope that they would make home. But they ran out of petrol and were forced down off Thorney Island, Sussex. When the aircraft hit the water Ercolani was injured and he went down with the sinking bomber, but the cockpit section floated to the surface, allowing him to join his crew in the dinghy, which then floated into the North Sea and eventually along the English Channel. The six crew drifted for 57 hours in their dinghy before they were washed ashore near Ventnor. Flying Officer Ercolani was awarded an immediate DSO, a very rare accolade for so junior an officer, 'for outstanding courage, initiative and devotion to duty.' [49]

Evidence shows that the enemy opposition was light. Of thirteen Halifaxes sent on minelaying near Oslo, three were shot down. Of the aircraft which returned safely, only 13 were damaged by shells; a sure indication that flak was generally light. No night fighters were reported. There was only one feasible explanation; most of the aircraft which did not return had flown round the bad weather and used up their fuel before reaching home. The most serious losses were those sustained by the Whitleys of 4 Group and the Wellingtons of Nos. 1 and 3 Groups. Ten out of 54 Whitleys failed to return and 19 out of 161 Wellingtons. Ironically however, the highest percentage loss of the night was suffered by the Hampdens of 5 Group which had been diverted from Berlin to Oslo and the Ruhr. Here the losses were five out of 19. It was a particularly black night for the Whitley which, though still classed as a 'heavy,' did not have the same range as the other aircraft out that night. Indeed, the raid of 7 November proved to be practically the aircraft's swan-song. Five months later the Whitley was at last retired from Bomber Command, though it was to continue in service long afterwards for Coastal. The Hampden followed the Whitley into retirement the following September. Alone amongst the 'old guard' the well-beloved 'Wimpy' was destined to continue into battle for many long months to come.

That this raid was unusually costly is not in doubt, nor are the factors which contributed to the losses: the extraordinarily bad weather; the limited range of the aircraft then available; and the comparative inexperience at that time of mounting such large-scale bombing attacks. The losses were referred to by Winston Churchill as 'most grievous' and in a note to the Secretary of State for Air and the Chief of the Air Staff after the raid he said, 'There is no particular point at this time in bombing Berlin... There is no need to fight the weather and the enemy at the same time.'

Early the following year 'Gee' navigation, which necessitated targets at shorter range, came into operation and the raid on 7/8 November was the last to be made on the German capital for over a year. When the attack was remounted, the aircraft used were Stirlings, Halifaxes and the immortal Lancaster. Aircraft which could much better withstand both the weather and the enemy defences. In the years to follow, the men who flew these famous aircraft won many laurels. But those others who bombed in the early years and laid the foundations for mass onslaught on Germany, must not be forgotten. They flew and fought with the aircraft then available and were not deterred by the limitations of these aircraft in extreme conditions.

The public learned something of what had happened from the BBC news bulletin at nine o'clock the following evening. 'Reports are now available from our aircrews,' read the announcer, Joseph MacLeod, "on last night's freak weather which interfered with the biggest offensive our bombers have yet launched against Germany... It was not only phenomenal weather, it was unexpected...' In one respect, however, this report was not entirely accurate. The bad weather was not unexpected. The met officers had known what conditions were likely to be as early as the morning of the 7th. Thirty-seven aircraft in all had failed to return. Although this was Bomber Command's heaviest loss to that date, it was only about two per cent greater than the average losses of the period. But what must have worried the planners at

Bomber Command was the staring fact that most of these aircraft were not lost through enemy action. They had not been shot down by night-fighters, nor by flak. They had run out of petrol. Over a pitch-black and cloud-covered Continent and over an icy, misty North Sea, 25 bombers had glided silently in to crash. Others were practically within sight of the British airfields when their engines stopped and an immediate crash-landing was the only chance left. And there were hundreds of telephone calls between airfields as other crews - the luckier ones - phoned up their bases to say they had just managed to scrape in at a coastal airfield.

The loss rate, which was no doubt a result of many aircraft suffering from icing or fuel starvation over the North Sea, led to a disagreement between Bomber Command and the Air Ministry and resulted in a conservation policy directive. For the foreseeable future generally, only weakly defended targets would be bombed. This would be the last major raid on the Big City until January 1943. [50]

Endnotes Chapter 6

41   *The Bomber Battle For Berlin* by John Searby (Guild Publishing 1991).

42   Just over 90 more bomber crews were given targets as far afield as the Essen and other areas, Ostend, Boulogne and Oslo, where 13 Halifaxes sowed mines.

43   *The Bomber Battle For Berlin* by John Searby (Guild Publishing 1991).

44   See *The Night The Fuel Ran Out* by Geoffrey Norris, writing in RAF Flying Review, October 1958.

45   *The Bomber Battle For Berlin* by John Searby (Guild Publishing 1991).

46   *The Bomber Battle For Berlin* by John Searby (Guild Publishing 1991). Sergeant A. L. Hassan RCAF and his crew were killed. Later, Searby and Fauquier were both Squadron Commanders in Bennett's Pathfinder Force. On 7/8 August 1943 one of the Path Finders who led the way for the Main Force was Wing Commander John Searby of 83 Squadron who was the Master Bomber or the 'Master of Ceremonies' at Turin. This was a trial in preparation for the role Searby would fulfil in the successful raid on Peenemünde later that same month. Before joining the Path Finders, Searby had commanded 106 Squadron at Syerston where he had succeeded Wing Commander Guy Gibson when he left to raise 617 Squadron. Wing Commander 'Johnny' Fauquier DSO** DFC won further fame when he stepped down from the prized rank of Air Commodore to take command of 617 'Dam Busters' Squadron, from Wing Commander James 'Willie' Tait DSO* DFC.

47   Williams, Sergeant D. J. Gordon, navigator; Flight Sergeant Donkin, wireless operator; Flight Sergeant J. R. F. Bourgeau RCAF and Sergeants R James and W Langhorn, the two gunners, were interned in Sage War Cemetery.

48   Abdon Ercolani was born in St Angelo, Tuscany in Italy on 8 May 1888. A picture frame maker, he migrated to London in search of work and in 1898 was joined by his family. He married in 1915 and he took British citizenship in 1923.

49   In October 1942 Ercolani left for India, joining 99 Squadron near Calcutta. The squadron was one of two Wellington long-range bomber units used to attack enemy airfields and river, road and rail supply routes. Ercolani led many of these missions over the ensuing months before the squadron switched to night bombing. Inadequate maps, appalling weather and poor aircraft serviceability due to lack of spares added to the hazards of flying during the 'Forgotten War'. With the expansion of the strategic bomber force and the introduction of the long-range Liberator, in September 1943 Ercolani went to the newly-formed 355 Squadron. He flew many sorties deep into enemy territory, some involving a round trip of 2,000 miles, to destroy the supply networks used to reinforce and support the Burma battlefield. An important and frequent target was the Siam-Burma railway built by Allied PoWs. In September 1944 Ercolani returned as CO to 99 Squadron, where he won the respect and affection of his airmen, who affectionately dubbed him 'The 'Erk'. He led many of the most difficult raids himself, often taking his heavy four-engine bomber as low as 100 feet to drop his delay-fused

bombs as his gunners strafed buildings or rolling stock. He attacked supply dumps and Japanese headquarters, and throughout the early months of 1945 regularly led forces of up to 24 Liberators against targets in Siam, southern Burma and on the Kra Isthmus, often in the face of heavy anti-aircraft fire. He was the master bomber for an attack against the railway system at Bangkok and was mentioned in despatches. By the end of March 1945 the decisive battle for central Burma was won. For his outstanding leadership and courage he was awarded a Bar to his DSO. Ercolani was then put in command of 159 Squadron, part of the Path Finder Force, attacking targets in Malaya and flying a number of mining operations to distant ports, including Singapore - sorties of more than 20 hours duration. On 15 June he led a force of Liberators to attack a 10,000-ton tanker, the Tohu Maru, which had been located in the South China Sea. The mission involved a round trip of 2,500 miles. Flying in appalling weather, some of the Liberators were unable to find the target, while some were damaged by enemy fire. Ercolani attacked at low level and made three separate bombing runs, registering successful hits on the tanker, which caught fire. Subsequent reconnaissance reports confirmed that it had sunk, a devastating blow to the Japanese troops depending on its vital cargo of fuel. Ercolani was awarded an immediate DFC. He flew his last operation on 5 August when he attacked a target in Siam. Almost immediately, his squadron then turned its attention to dropping food and medical supplies to the many PoW camps spread across Siam and the East Indies. Ercolani left the RAF in March 1946 and rejoined his father at Ercol. Owing to the scarcity of raw materials, new furniture had been rationed since 1942 and the particular achievement of the Ercolanis was to mass-produce the Windsor chair while conforming to the stringent requirements of cost and material laid down by the Board of Trade. At its peak of production, Ercol made around 3,000 Windsor chairs a week. For many years Ercolani served as chairman and joint managing director with his brother. He formally retired in the mid-1990s but remained closely involved with the company until his death on 13 February 2010. Lucian Ercolani married, in 1941, Cynthia Douglas. She died in 2004 and he is survived by their daughter; a son predeceased him.

50   Sir Richard Peirse became AOC-in-C India on 6 March 1942. On 16 November 1943 Peirse was appointed Allied C-in-C, Air South-East Asia Command. While serving in this capacity he eloped with the wife of General Sir Claude Auchinleck, C-in-C India, an act which was universally condemned and effectively ended his career. Peirse died in 1970 aged 78. *The Right of the Line* by John Terraine (Hodder & Stoughton 1985).

Chapter 7

Flames, Flares and Fires

'There is no question of sitting or lying in an air-raid shelter reading a paper or playing games and looking round about every half-hour - a terror raid is different from that. The air is full of the drone of motors, of the incessant roar of the flak, of the whistle and burst of bombs; night has become day owing to the flames and flares; fires rage in the streets, the heat becomes unbearable, brick dust makes breathing difficult and gets in the eyes, in the throat and in the lungs; over everything there is a biting smoke... Only he who has strong nerves, is blessed with courage and toughness, is a real man, can overcome these horrors. Useful help can be given only by those who are morally equal to the horror. Therefore it is necessary that everybody should accustom himself to the idea of what such a terror raid looks like and that everyone by sternly taking himself in hand should steel his nerves in order to pass the ordeal in the hour of trial.'
An article on ARP in a German newspaper.

A new British directive calling for 'area bombing' of German cities had been sent to Bomber Command seven days before Harris' appointment (though for the rest of the war he became synonymous with the attacks on German cities). The Air Ministry decided that bombing the most densely built up areas would produce such dislocation and breakdown in civilian morale that the German home front would collapse. With the new directive bomber operations at night entered a new phase that was not restricted to one side of the divide. On the night of 3/4 March 1942 the weather was clear and there was a full moon. A force of 235 bombers - the greatest so far to a single target went out, but not to any German target. The Renault works at Billancourt near Paris were bombed. There were few defences, the weather was perfect, bombing was at the lowest level at which the aircraft could be safe from the explosion of their own bombs and the huge factory was obliterated. The Renault works were making tanks and motor vehicles for the German army; the destruction meant that the Germans had been deprived in two hours and at a cost of two RAF bombers, of the equivalent of the whole armoured equipment and transport of five motorized divisions.

As March continued, the Ruhr and Kiel were repeatedly attacked whenever the weather allowed. Some significant but not spectacular damage was achieved. The *Gneisenau* had to go to Gdynia because Kiel became too hot for her;[51] the *Scharnhorst* appeared to have been damaged again. And in the Ruhr a quite small number of aircraft, in the course of an attack by a strong force on various objectives in that district, damaged the Thyssen steelworks at

Hamborn in the extreme west of the Ruhr. This was the largest steelworks not only in Germany but in all German occupied or German-controlled Europe; Krupps at Essen was a larger industrial plant, but Krupps turned out finished armaments as well as steel and pig-iron. The crews who attacked the Thyssen works in March took photographs as the bombs went down and it was clear from these photographs that there were fires in the area of the works; in one photograph the burst of a very heavy bomb could be seen within the target area. The Ruhr was a labyrinth of industrial buildings and the crews could not have been positive which particular factory they had bombed without the evidence of the photographs. This damage had not been completely repaired in July when the Thyssen works were hit again and even harder and when another reconnaissance of the area was made.

A few nights after Lübeck, on the night of 1/2 April, 35 Wellingtons and 14 Hampdens carried out low-level attacks on railway targets at Hanau and Poissy; a new experiment which proved to be too costly. Bill Anderson recalled: 'They sent the squadron on another change of target, low-level bombing of railway lines and stations near Frankfurt and we rubbed our hands with real pleasure. It would be something new and it would be away from Happy Valley. The weather on the way out was bad and we ran into electrical storms. The propellers were ringed with blue fire and the wing tips bathed in an eerie glow of pale mauve. Little fat sparks were jumping off odd fittings. And it was very bumpy, so bumpy that on one occasion George temporarily lost control. However, we ploughed on and gradually the weather cleared and we could see the ground. We found the river we were looking for and that led us to the railway. We flew up and down the line dropping bombs on the little wayside stations, hoping to find a train that we could shoot up. However, they all seemed to be hiding in the tunnels, so we turned for home.

'One of the games of navigation that has always fascinated me is finding position by means of the stars. I tried a little on the way home just for practice when we were well away from the enemy defences and to my disappointment, my results were very bad; I was eighty miles farther south than I knew was possible, though since the ground below was covered with cloud, I could not be certain of my exact position. I tried again, wondering why I had been so far out. This time I was a hundred miles farther south than I ought to have been. Feverishly I checked up everything. One or two things must have happened. Either the stars had moved from their courses, or else the wind, which was about fifty miles an hour from the west on the way out, was now about a hundred from the north-west, I decided that the latter was the more likely! So we turned and butted our way slowly back towards England. Two and a half hours late, we crossed the enemy coast and they had all but given us up when we finally got back to base. We landed and went to interrogation.

'When you return from an operation, the first thing that happens, even before interrogation, is that you report to the squadron commander, who grins cheerfully and asks how things have gone, you tell him briefly what has happened, so that he can get a picture of the operation. This time, however, [Wing Commander Leslie Vidal 'Jimmy' James DFC] the squadron commander was not there, it was the navigation officer with a curiously set face and a

'Thank God you are back'. The squadron commander was in the control tower listening desperately for news of some of his lost aircraft. For only four of our squadron got back that night, though one more landed later in the South of England.[52] It was the weather that lost us those crews, not the efforts of the Hun. There are two major problems with regard to weather. They are wind and icing. That night the losses were due to the wind. As we were coming in to land, the squadron commander was listening to one of his friends calling for help somewhere near Happy Valley, where he had been blown by the wind. But icing was perhaps the worse danger. A white film would build up on the front edge of the wing and the windscreen would frost up. Lumps of ice would form even on the propellers and fly off to smack against the side of the fuselage. Luckily, severe icing conditions were fairly rare and as the war went on and aircraft heights went up from ten to twenty thousand feet, it became a less common source of danger. For up there, moisture in the air is like very finely powdered snow, too fine and too cold to make a snowball of the aircraft. Though sometimes as you breathe, flakes of snow patter against the back of your throat.

'Back to Happy Valley we went next night, where the guns were ready and waiting. Where smoke and mist covered the ground so that we never could be certain that we had found that which we were looking for, where so many of our friends had been caught like great moths in the searchlights and like moths had burnt and fallen. And George was getting jittery and jumpy. That night we lost a third of what was left of the squadron and the mess afterwards was very quiet.

'A couple of mornings later, the squadron commander called me into his office. He told me in that curiously soft voice that sounded so odd coming from such a tremendous figure, that he was going to fly himself that night and I was to navigate for him. He thought the squadron was a little depressed and he intended to show them that things were not as bad as they might seem. We would make three runs over the target and bring back photographs. It would mean flying rather lower than might be pleasant, but I was to be very careful only to bomb if I was quite certain of hitting the right spot, otherwise we would go round and try again. That night he flew out quietly to the target and did exactly as he had planned, just as if it was a practice exercise and the defences did not exist. The target was Essen.

'By all the laws of justice this wonderful example should have been a real tonic to the squadron, he made it all look so ridiculously easy. Nobody who operated with Wing Commander 'Jimmy' James could help being affected by the utter simplicity of mind and singleness of purpose with which he attacked a target. To say that he took risks is beside the point, he apparently never considered them. He had a load of bombs aboard which he was going to deliver to a German address and no power on earth, least of all the German defences, was going to stop him.

'I flew with Wing Commander 'Jimmy' James twice more. First to Rostock. Owing to a 'boob' on my part - I mistook one of the islands in the Baltic - we were late on to the target. However, the Wing Commander decided to drop the bombs two at a time, so he made a few runs low down amongst the light

152

flak. We came in for a good deal of individual attention which seemed to tickle Wing Commander James and when he had quite finished we set out for home. Just as we came to the Danish coast, the second pilot noticed that the petrol was nearly finished, we must have been holed. We had enough to make Sweden, not enough to get us back to England.

'Wing Commander 'Jimmy' James had arranged to go to Rostock, bomb it and then come back. So he set out for England. He brought the aircraft low over the water to keep out of the strong head winds that were blowing upstairs. Then he handed over to the second pilot, clambered past me and sat down by the reserve taps ready for when the engines would run out of juice. I was not frightened, just furiously angry with myself that through my getting lost on the way out we had wasted those all important gallons. From where I was working I could see Wing Commander James sitting there waiting calmly and it hurt. Wearing a helmet and a flying-suit, he seemed no longer a 'Wingco' in charge of a squadron, but just a man, completely in command of himself and the situation. Every now and then he would turn and smile and hold his thumb up. On and on we droned towards England. I worked out the point at which our reserve tanks would be enough to get us to land and as we got nearer, I began to hope.

'The minutes tickled slowly by until at last we reached the point where I knew we could come in on our reserve tanks. I called Wing Commander James up and told him we had enough petrol to get us home. His reply was characteristic.' Of course we have Andy, you silly chump, it was only the petrol gauges reading wrong.' I felt an awful fool until a thought occurred to me; he could not have known this when he set out from the Danish coast!

'The next time I flew with Wing Commander 'Jimmy' James was to Essen again. He never seemed to choose the nice, easy trips. He made a couple of runs over the target and on the way back attacked a German night fighter aerodrome with guns and got chased away pretty firmly by some light flak. I remember thinking at the time that he would not live long at this rate. Yet I always felt safe with him and I would have hated it if he had asked anyone else to navigate him.

'The next time Wing Commander James operated [to Emden on the night of 20/21 June] I was not there to go with him. He did not come back. A Hun fighter got him on the Dutch border and he was killed.[53] If only George had operated with Wing Commander James all might have been well with him, but unfortunately we had taken another man as second pilot. And when we went to Happy Valley the next time, George found that something was wrong with the oil pressure and we had to turn back before we crossed the enemy coast. But it must all have been in George's imagination for nothing wrong could be found next day. Then George pulled himself together. Next night it was the Ruhr again, but he was his old self once more. A cheerful type who cared for nothing or nobody and we flew as a crew and not as six seated men.

'Next it was Hamburg, not such a long drawn-out battle as Essen, but possibly even more intense over the target. Everybody has prejudices against certain places and I think Hamburg was the one that I personally liked the least. The guns at Essen were many, but somehow one felt that they were fired at the aircraft and not at you personally. The guns at Berlin were frantic, trying to keep you away from the capital of the Reich, but the guns at Hamburg were downright vicious. They fired to kill; besides, I had lost my best friend on the previous trip to Hamburg. He did not want to fly that night 'because his mind was somewhere in Worcester where his wife was very ill. But excuses like that can't be made unless the target is an easy one. If it's one of the warmer ones, you just cannot stand down. He went and he did not come back.

'Yet on this particular night when we set out for Hamburg, the crew were in cheerful form. No longer was the skipper shaky, no longer did he worry about little things that did not matter. Once more he was his old self. And we were taking with us as second pilot, a new flight commander, a man with many hours flying but as yet no operational experience. We were to show him what a target in Germany was like.

'We crossed the North Sea and turned down the north side of the river Elbe to Hamburg. It was all very quiet and a little sinister. George was joking cheerfully, Suddenly Hamburg defences opened up just ahead and a wall of searchlights with the flak bursting in little tight clumps. George said in a most melodramatic voice, 'I can't go in there' and we were rather tickled. What cracking good form he was in! A few minutes later, the well-known crunching sound of near flak bursts began. I drew my feet a little tighter under my seat and made that last check as to the exact time that we ought to be over the aiming point. Then I collected the photograph of Mary, unpinning it carefully from the wall and began to stuff it down inside my Mae West, ready for that crawl forward to the bomb-aimer's position. At the same time I felt that the aircraft was flying a little oddly. I never finished tucking the photo away.

'The door leading forward to the pilot's cockpit suddenly opened and for a moment I saw a figure silhouetted against the beam of a passing searchlight. Next moment, George was sprawling across me, unconscious. And his body pulled my intercom plug out. For a few seconds that seemed like ages I fumbled before I could plug in again and instinctively I called Eddie. Back came his voice at once, 'OK Andy, second pilot has taken over.' And indeed, the second pilot, being new to operations was probably the least staggered member of the crew. He simply watched the whole proceeding with attention and then climbed into the vacant seat and drove on as if this was all part of the sort of thing that happened on trips over Germany.

'Eddie helped me and together we got George on to the bed. There he lay quite unconscious. I smoothed out the crumpled photograph of Mary and went forward to do the job. Ten minutes later we were headed for home and George was coming round, mumbling over and over again to himself, ' Andy won't fly with me again Andy won't fly with me again'.

'And I didn't. George never operated again. The RAF went into his case with their usual thoroughness and he was permanently grounded and eventually sent back home.

'This cracking up was fortunately rare in Bomber Command. And in most cases, the trouble was only temporary. There was the gunner who went to his squadron commander and told him quite frankly he was too scared to fly any more. Now this gunner had been shot down once already and yet had carried on, so his squadron commander, in order to give him a rest off operations and yet be able to keep an eye on him, gave him temporarily the job of helping on the crash tender. On every airfield they have one or two crash tenders standing by ready to rush to any aircraft that crashes, in the hope of being in time to deal with the ever-present danger of fire. So naturally, when they have to go anywhere, they have to go in a hurry. They were having a practice run one afternoon and the gunner, who was sitting on top, fell off and fractured his skull as well as cracking a couple of ribs.

'Some weeks later, while the luckless gunner was convalescing, the squadron commander paid him a visit and the gunner at once began to clamour to get back on to operations. The squadron commander, thinking this was just one result of the crack on his head; tactfully put him off. However, the request was repeated even more urgently on subsequent visits and at last the squadron commander thought it advisable to remind him of his previous unwillingness to fly. It transpired then that it was not the blow on the head that had upset the gunner's memory, but merely that the period of convalescence had enabled him to review the whole situation. He had, therefore, come to the inevitable conclusion that a nice, quiet operation against the Hun was preferable to the perils of life on the crash tender.

'Yet most people overcome their fear in a less spectacular way. There seem to be two stages. First, the realization of one's own cowardice. It is an unpleasant realization, to be hidden away from the rest of the world, but it has to be faced by yourself. Next comes the comforting discovery that even though frightened, it is still possible to carry on and do the job. And from this second discovery grows a confidence that cannot be shaken, for it is founded on a knowledge of your own weakness. For you know now that just as you have carried on in the past with a dry mouth and an empty feeling in the pit of your stomach, so you will be able to carry on in the future with a dry mouth and an empty feeling in the pit of your stomach.

'And fear is a wonderful medicine. It purges man clean of greed and selfishness and all the little meanness's that make him so miserable. So that In the darkness of the night, with the stars above and death below, he is lifted for a few hours by fear and awe up out of the struggle for petty advantages that is such a large part of our life on the earth.

'Air crew are not saints, even less so than most men. But like all men, they have inside them a longing for better things. And it is this exultation in the face of danger, this purification by the virtue of fear that lies behind their contradictory statement, 'I'm bloody well scared stiff the whole time and yet there's something about being on ops.

'For Bomber Command, the first two years of the war were years of air raids. Indeed, during the 'phoney' war winter of 1939, many of the raids were only to drop leaflets. There is the tale of the Wellington that was flying over Berlin dropping out inflammatory literature. The natives were objecting pretty strenuously and splinters of flak were rattling against the fuselage. The wireless operator began to get a little bored with scattering leaflets so slowly and dropped a few packets out intact. Unluckily, the skipper saw him and was horrified. 'Good heavens, man, you mustn't do that and they might hurt somebody!'

'The tales of these early raids make curious reading in the light of the mass attacks to follow. For those were the days when fifty Whitleys constituted a major effort, when the target was the post office in the Wilhelmstrasse and the time was Friday night. You went your own sweet way and you chose your own good time. Things were very sporting and not very highly organized. Briefing was brief, really brief, the record being five words from a squadron commander, 'St. Omer, chaps, follow me' and clutching his maps firmly in his right hand, he clambered into the cockpit of his Blenheim. But this attitude did not always produce the expected results. There is even one highly apocryphal story of a navigator whose geography was a little weak, who one night was briefed to go to Hamburg. He hunted all over his map and at last found the place, though curiously enough an abbreviation was written against it, instead of the whole place name. He went to the target and they shot at him pretty sternly and he came back rather cocky because as far as he could tell, he was the only one who got there] At least, he had not seen anyone else getting a pasting. He was asked if he had seen the docks. Docks? No, there weren't any docks there, only masses of railway lines. He had carried out a solo attack on Hamm!

'The material damage done to the Hun during those first two years may not have been very great, but the effect on his morale must have been pretty considerable. For he had so often been promised by Goebbels that he never would be raided, that he must have taken a pretty dim view of these efforts of Bomber Command, spasmodic and sporadic though they often were. And then gradually the air raids began to change their character and became air attacks, carefully planned operations designed to hit the Hun in places where it would hurt the most and in enough force to swamp his ARP services. Four-engined bombers carrying tremendous loads came into service. Times on targets and routes were laid down precisely. New navigation devices were introduced which had the effect not of making navigation less work, but of making it more certain. Attacks were carried out on the Hun industries even when they were covered by cloud or hidden in smoke and haze. And most important of all, the results of individual crews began to be assessed by night photography.

'Air crew are optimistic by nature, otherwise they would hardly be air crew. And they hate to feel that a sortie over enemy territory in the face of flak and fighters has been wasted by the bombs being dropped in the wrong place. The wish is father to the thought. The hope that they have hit the right place becomes the belief that they have. So that the results of night photography, which enabled crews actually to photograph the place that they had hit, came as rather a shock. Those that were most positive that they had identified the post office in the Wilhelmstrasse, returned with pictures of a distinctly pastoral flavour. It became obvious that gallantry and good intentions were not enough. The ideal aircrew was not the glamorous knight errant seeking for adventure, but rather the careful and accurate technician, following precisely a well-laid plan and eliminating all unnecessary, hazards. No longer was the demand for a Galahad of the skies wielding a flaming word; the war now called for a competent craftsman, working a simple slide-rule.

'April and May were quiet months for the squadron. It was as if Bomber Command was saving up for something. A few mine-laying operations were laid on and being a change, they were as good as a rest cure. One beautiful clear night we flew out low over the North Sea. A red glow appeared ahead of' us which turned out to be our old enemy the moon. For though she is lovely, she is false and her beauty is cold and fatal. She shines on the woolly clouds and makes them soft and white, but silhouettes the black form of your aircraft for the fighters. However, we sneaked safely across Denmark and down the Baltic to the place where the mines were to be laid. As bad luck would have it, we flew low over a ship and as soon as we were safely past, they let fly at us with cannon. I was standing by the captain at the time and our hands were going from side to side after the ludicrous fashion of Wimbledon crowds, as we watched the shells spraying past. There was a bang and the aircraft began to fly a little oddly and a check round the crew brought no reply from the tail gunner. A shell had hit the tail and bent it, broken the Hide of the rear turret and cut the intercom. Worst of all, it had punctured the luckless rear gunner in a part of his anatomy favoured by tradition for corporal punishment. We got him out and made him comfortable. Then arose an argument as to how the incident ought to be recorded in the log. For some held that, as a gunner facing aft, it was his starboard buttock that had been smitten, while others argued that as an individual it was his port buttock. The discussion was lively on the part of all but the victim, who appeared to consider that the argument was a trifle academic, though he did not put it so politely. They gave the skipper a DFC for this trip.

'Towards the end of May, rumours began to float round the squadron. It was obvious that something was brewing. To begin with, the intelligence 'types' had walked about for days looking very innocent and changing the subject rather obviously whenever the question of targets next on the list was raised. There was much hurried rustling of papers being crammed into drawers when one paid them the odd social call. Altogether they behaved like small boys who had newly joined some secret society. The ground crews, too, were busy getting all the available aircraft on top line. On the few days when the squadron was on operations, full strength was not called. Obviously they were saving up for something. And finally the Squadron CO and the Station Commander would disappear at intervals and have long and mysterious interviews with the powers at Group Headquarters. Everything pointed to a really big show in the offing and the general opinion favoured a low-level attack on Hamburg.

'One morning the excitement and general 'flap' reached a crescendo and we learnt that we were all 'on' that night. We smelt that something big was in the air and it was with a certain anticipation that we crowded into the briefing room. The anticlimax was awful. We found it was just another trip to Happy Valley, this time to the city of Cologne.'

'We listened carefully to the briefing but without very much enthusiasm. We felt somehow that we had been cheated. The elephant of rumour had given birth to a miserable mouse. The route was straightforward and the boys knew it well. 'Be careful of Amsterdam' - they were liable to take a crack at you with those batteries of heavy flak round the town. 'Don't get too far north on the way in and so run into Mönchengladbach' - tradition had it that there was a

German anti-aircraft school there and the instructors would rush out and give their admiring pupils a demonstration of how it ought to be done. 'All round the target itself there will be plenty of searchlights and more stretching up northwards towards Düsseldorf' - most of us knew this from bitter experience. 'The town is an important centre, the second largest city in the Reich with a population of...' we had heard this line of 'Bull' so often before. 'Our job is to carry incendiaries to start up fires for the other crews to bomb' – same old job in the same old Happy Valley. Why couldn't they think up something new for a change; after all the fuss and gefluffle of the last few days, we were looking for something really big. 'The weather is expected to be clear over the target' - we had been disappointed too often in the past by the Ruhr mists to hope for very much, for it always seemed to be hazy in that part of the world. 'Base conditions on return should be good' - thank heavens for that at least, it is pretty grim to come back after a long trip and find that you can't get down at your own place. You then have to hunt round the countryside for some god-forsaken aerodrome that no one has ever heard of and spend the few hours before dawn in armchairs in the mess, for the beds are sure to be full up.

'One more thing. There will be more than a thousand aircraft on the target.'[54]

'The climax was terrific and in a moment everyone was cheering and laughing. So this was IT. This was what we had been waiting for so long. The Hun had asked for it. Rotterdam, Coventry, London and now, by God, he was going to get it.

'Every aircraft that the squadron could muster was there at take-off and they followed each other lurching round the airfield, a long line of giant, roaring beasts. Then one by one they swung on to the end of the runway and then went roaring and pounding off into the night.

'The sky was clear above, the ground was clear below and the coast of England stood out black against the sea. On the port beam trembled the Northern Lights, a glorious panorama of shifting colours. All the stars were out to greet us and as we turned southwards towards Cologne we could see flying along below to port, above and to starboard, other aircraft all pressing steadily onwards to the target. Here and there over the Dutch coast, a searchlight wavered and a few guns fired. From the farmhouses in Holland, the V for Victory sign was flickered out to us. Soon we were into Germany and as we came up to the Ruhr, the defences sprang into life. A couple of dummies, areas of lights that twinkled like incendiaries on a target, complete with canopies of searchlights, were hopefully lit up to attract attention, but we thumbed our noses at them and passed on, We were out to get Cologne that night. Ahead were the defences of the city, below and to the left, the River Rhine was shining like a strip of silver ribbon. Great cones of searchlights were building up and as we came in, the flak opened up and at the same time the first incendiaries began to fall. Half-way through our bombing run, the searchlights caught us and blinded us and a burst of flak holed us amidships. We turned off and came in again, so that by the time that we finally bombed, the 'party' was in full swing. The defences were hopelessly swamped. The searchlights were waving helplessly and many of them had gone out. All the guns seemed to have stopped firing except for one battery in the south of the town, which, ringed with glowing red fires, still blazed hopelessly away into the skies. Amongst the blinding white incendiaries and billows of smoke, lights were leaping as bombs fell amongst the burning buildings. And every now and then, an especially bright flash would be followed by a huge, slow mushroom of smoke as a 'cookie,' two tons of concentrated destruction, burst in the doomed city. And as we left the target, we could see behind us a great pall of smoke building up over the heart of Cologne.'

Sergeant Jon Rootes on 149 Squadron described his experience of Cologne thus: 'The silver ribbon which was the Rhine River gleamed in the bright moonlight on the starboard side. Wellington X9617, an old hack bomber off 'circuits and bumps' at 21 OTU at Moreton-in-Marsh, was in the foremost wave consisting of Wellingtons from 3 Group - the first wave of over 1,000 bombers on their way to raid the great University city and port of Cologne. I was still feeling the thrill generated at the briefing that afternoon. We had known, of course, that the 'big one' was in the offing because there had been no flying for three weeks and Tiger Moths were constantly landing with replacement aircraft parts. Over one thousand aircraft would be flying that night. They were mostly twin-engined planes and included 300 from training and conversion units. Excitement ran high as the battle plan unfolded. At last, after many frustrating raids in bad weather with little damage done to the enemy, here was a chance to prove that we could do it better - and hopefully shorten this war.

'Some of the crews were extremely alarmed by the prospect of a collision in such a large formation. They were assured by the Bomber Command's operation research scientists who had computed that there should be no more than two aircraft colliding over the target! What assurance! And they weren't flying, were they? 'Okay,' said one wag. 'Do they know which two?'

'Then they had walked out into the dusk of a fine evening to take off at 22.55 hours through clear skies towards Cologne. Sergeant Rootes had never experienced so many aircraft so close - and at night! One had flown right over his aircraft, just like a giant bat, missing the tailfin and flying on apparently completely unaware of the near-miss! The trip to Cologne was uneventful and his Wellington dropped its bombs on the target and turned for home. Rootes looked down from his gun turret to witness an incredible sight. The great city was plain to see in the vivid moonlight. The twin towers of Cologne Cathedral stood tall and untouched amidst the rubble around it and the Rhine Bridge. The centre of the city was already a core of fire; searchlight beams stroked the sky as if drunk. The blazing incendiaries looked like a carpet of diamonds winking up at him. He had never seen such devastation. In many raids and many fires, he would never forget Cologne.

'On the trip home the Junkers night-fighters were busy. Sergeant Rootes counted five Wellingtons ablaze and falling. One of them was directly astern of him. The port propellers slowed as it was feathered but the flames grew higher. He knew that the fighters would come in again and their own aircraft would be illuminated by the flames of the stricken Wellington astern. He watched as the fighter's cannon continued to knock pieces out of his victim. The pilot of his prey was throttling back and his fuel was atomizing as it left the trailing edge of the wing and ignited like a gas flame. Lumps fell off the aircraft, but no parachutes showed. He shut his eyes tight as the remains of the Wellington slipped below him. Wellington X9617 got safely home to land five and a half hours after taking off. Forty three others did not return, one hundred and sixteen crash-landed or arrived home badly damaged.

'He remembered the telegram he had received whilst on leave. Was it only yesterday? It had simply stated: 'Return to Unit'. He hoped he always would![55]

'The 'thousand' raid on Cologne certainly touched the public imagination' recalled Bill Anderson. 'They saw it as a tremendous blow at the Hun and as a proof of the growing might of the Royal Air Force. But to us, it was something more significant than that. It was the outward and visible sign of a great new step forward in the bomber offensive. It stood for the end of those casual raiding days and the beginning of an era of calculated hammer blows from the air, based on precise planning and careful attention to detail, planning that would mean that henceforth every crew that went down in flames over Germany would be paid for over and over again by the Hun in his burning and bleeding cities.[56]

'By comparison, the next raid was rather an anticlimax, for though it was another 'thousand' effort, the weather was bad. The target was Essen, but we had a peaceful ride.[57] Incidentally, it was on this trip that I first saw a 'scarecrow.' This was a weird and wonderful contraption fired up by the Germans which burst in the air and floated for a time with streamers of flame dripping off it. It was supposed to look like an aircraft on fire. They then tried to complete the effect by coning it with searchlights and firing off at it like nobody's business. This was supposed to be extremely frightening to the British Air Huns.

'I did one more trip to Essen and had another very peaceful evening. Happy Valley was like that, either you had a hot time or else some other luckless type nearby got the 'pasting' and you went through scot-free. Two days afterwards I was sent to Bomber Command Headquarters to work on the Navigation Staff. Another navigator took my place with the crew. A few trips later, they were flying home peacefully when a fighter came up from below and raked them from end to end. It was Cyril, the Yorkshire front gunner, who told me about it. He was in hospital with a broken arm and holes in his body. The skipper had managed to get the shattered aircraft back to England and then crashed her, for she was too badly knocked about for her undercart to work properly. He was afraid she might catch fire, for there was petrol about and the engine exhaust pipes were red-hot and not realizing Cyril was badly wounded, he hauled him quickly out of his turret, which hurt pretty badly. Cyril told me all about this as if it was rather a good joke and it was only when I asked about Eddie that he lost his cheerfulness. He told me that Eddie had not been hurt. A few days later he was back on the squadron. A new crew were short of a wireless operator, so Eddie badgered to be allowed to go with them. It was a hot target. They didn't come back.

'Eddie was a very ordinary man with a wife in Wellingborough. A man who did his duty because he was made that way. A man who carried on because it was not in him to do anything less. Once when he insisted on going on a rather 'shaky do', I asked him why he wanted to stick his neck out, why he wanted to run into trouble. His reply was simple and typical. Two words: 'Why not?'

Endnotes Chapter 7

51   The *Gneisenau* was severely damaged along the whole length of her forecastle and needing extensive repairs. There the Russians found her on 28 March 1945. She had been partly dismantled, scuttled and used as a block ship long before the Red Army entered the port.

52   22 aircraft reported that they had carried out this task but 12 Wellingtons - 34.3 per cent of the Wellingtons dispatched - and one Hampden were lost. 57 Squadron at Feltwell lost 5 of the 12 Wellingtons it sent on this raid and 214 Squadron at Stradishall lost 7 of its 14 Wellingtons. *The Bomber Command War Diaries: An operational reference book 1939-1945* by Martin Middlebrook and Chris Everitt.

53   Wellington III X3713 WS- 'J for Jane' was lost with all five crew.

54   Harris could only accomplish 1,000 bomber raids by pitching in untried crews from the Operational Training Units (OTUs), many of them flying Wellingtons, but even Blenheims and Hudsons were used. When 1,046 aircraft were sent to bomb Cologne on 30/31 May, 599 were Wellingtons and no fewer than 367 of the aircraft came from OTUs. Not only was the attack planned to be over in 98 minutes; it actually was executed in that time. To quote from Air Marshal Sir Arthur Harris's message to all his crews on the eve of the attack, a force 'at least twice the size and with more than four times the carrying capacity of the largest air force ever before concentrated on one objective, reached Cologne, at the rate of one bomber every six seconds.

55   *The Night of A Thousand Planes* by Jon Rootes DFM quoted in *'Ordinary People'; True Accounts from wartime ex-aircrew* edited by Cyril Thompson (Barny Books 1999). His pilot, Squadron Leader The Honourable Brian Grimston DFC and crew on 156 Squadron at Warboys were killed when their Lancaster crashed in the target area on 4/5 April 1943. One of his fellow officers described The Honourable 'Grimmy' as 'being so tall that you felt he ought to wear oxygen when he was standing up'. 898 crews claimed to have reached and attacked the target. They dropped 1,455 tons of bombs, two-thirds of them incendiaries. More than 600 acres of the city were destroyed. In all, 40 bombers and two Intruders were lost, a 3.8 per cent loss rate; 116 were damaged, 12 so badly that they were written off. The fires burned for days and 59,100 people were made homeless.

56   At dawn on the morning of 31 May a reconnaissance aircraft took off in England and in an hour or two was back again with a report that Cologne was hidden under a cloud of smoke which extended to a great height. There could be no question at that time of making any photographic reconnaissance. The smoke cleared away in time and after some days an adequate survey was made of the city. The whole of it was eventually covered by good clear photographs. One-third of a city of 761,000 inhabitants had been laid waste. The whole centre of the city was gone, but miraculously the cathedral still stood. Of the 250 factory buildings destroyed, or heavily damaged, in all parts of the city, thirteen were important plants engaged in engineering, metal, machinery and foundry production. On the west bank of the Rhine four chemical plants, three rubber works, three factories making electrical apparatus, an oil-storage depot, a steel-rope factory and a textile works were destroyed or damaged. The great and extremely important Nippes railway workshops were put out of action altogether. On the east bank there was damage to both the two great Humboldt Deutz factories, one in the Kalk and the other in the Mulheim district, which besides much else make Diesel engines for submarines. Also on the east bank there were destroyed or badly damaged a factory making accumulators for submarines, a tyre factory, a plant producing rolling stock, a chemical works and a factory making undercarriages for aircraft. Three hundred acres in the commercial centre of the city was devastated and about 140,000 people, it is said, were officially evacuated. Trains left Cologne crammed with people - even goods trains were used. Great numbers left by road, walking as far as forty miles before they found shelter. Representatives of the Ministry of the Interior were sent to Cologne and passed thousands of these refugees on the road. They shouted at the party from the Ministry: 'You here, too! This is the bill for London that we are paying! ...This is revenge for Coventry!'

57   The second 'Thousand Bomber Raid' took place on the night of 1/2 June when 956 bombers including 347 from the OTUs went to Essen. 31 aircraft (3.2 per cent) failed to return. There was no wind on the night of 1/2 June and the haze was extremely troublesome. A third 'Thousand Bomber Raid' took place on the night of 25/26 June when 1,006 aircraft, including 102 Wellingtons of Coastal Command, attacked Bremen.

Wellingtons at dispersal
being bombed up.

The aftermath at 7 am on the morning of 30 April 1942 as taken by Flight Lieutenant P. N. Atkinson from his 278 Squadron Lysander IIIA at RAF Coltishall in Norfolk.

Man-handling water pumps and hoses, men of the National Fire Service in Norwich continue to fight fires after their vain attempt the night before to save some of the city's most famous stores.

Orford Place in the centre of Norwich on the morning after the raid on Wednesday 29/30 April when an estimated 75 enemy aircraft operated over Britain. Forty of these carried out a second devastating raid on Norwich in which the bombing was more concentrated on the city's commercial centre.

Devastation to Dereham Road and St. Benedict's (right) during the Baedeker bombing raids on Norwich.

Wellington N2912 on 215 Squadron, seen here at Bassingbourn in mid-1940. On 24 April 1941 this aircraft, now on the strength of 11 OTU, was shot down directly over Bassingbourn by an Intruder flown by Feldwebel Gieszubel. Completely out of control, the stricken Wimpy crashed directly onto another of the unit's aircraft (R1404) parked at dispersal. Sergeant Alstrom, the 18-year old pilot and 30-year old Sergeant Wilson were both killed although the third crew member, Sergeant Nicholls, walked from the wreckage with only minor injuries.

A Wellington crew at Mildenhall study the map prior to take off for another raid on Berlin.

Hampden EQ-C on 408 Squadron RCAF at a snow-covered dispersal at Balderton in January 1942. The Canadian squadron flew over 1,200 bombing sorties on Hampdens between June 1941 and September 1942.

Right: The son of an Italian furniture designer and manufacturer who had come to England in 1910, Pilot Officer Lucian Brett Ercolani on 214 Squadron who was awarded an immediate DSO, a very rare accolade for so junior an officer, 'for outstanding courage, initiative and devotion to duty' on the raid on Berlin on the night of 7/8 November 1941 when 21 aircraft including ten Wellingtons (including Ercolani's) were lost.

Left: On the night of 12/13 August 1940, Flight Lieutenant Roderick Learoyd was the pilot of a 49 Squadron Hampden, P4403/EA-M, one of eleven dispatched from Scampton to make a low-level attack on the aqueducts carrying the Dortmund-Ems Canal over the Rover Ems near Münster. Learoyd received the Victoria Cross. *(IWM)*

Wing Commander (later Group Captain) John Alexander 'Speedy' Powell OBE DSO commanding 149 Squadron. Powell had earned his nickname because of his obsession to get to grips with a problem as quickly as possible. The Powell motto was 'Never Fear', a term he used a great deal. After every briefing, the crews would inevitably be sent on their way to their target with the admonition from Powell: 'Don't do anything I wouldn't do!' - which in fact meant that they could do pretty well anything they wanted, as long as it got the job done! He was a man at ease with the most junior of the erks and the most officious brass and would offer his gold-plated snuffbox equally to both. 'Speedy' Powell (pictured here in the Middle East where he commanded a Wellington Wing) failed to return from an operation on 8 August, 1944 - some say flying an American P-38 and others say flying a Hurricane over Yugoslavia.

The Hampden in action - a vital aqueduct of the Dortund-Ems Canal being bombed by Flight Lieutenant Learoyd.

On 12/13 November 1940 77 Bomber Command aircraft were dispatched to attack a number of targets in the Ruhr, 102 Squadron seeking all oil refinery at Wesseling. Pilot Officer Geoffrey Leonard Cheshire and the crew of Whitley P5005/DY-N had some confusion in finding the target and after 20 minutes' searching Cheshire decided that they should bomb Cologne's marshalling yards instead. While investigating beneath the clouds, two direct hits were taken from light flak. One burst in the fuselage, detonating a flare and opening up a 15 feet section of the port side. The WOp, Sergeant A. Davidson sustained slight wounds and the interior of the Whitley was temporarily filled with acrid fumes, Cheshire momentarily losing control and only managing to regain it after the aircraft had lost 2,000 feet in a dive. Despite the adverse aerodynamic effect of the flapping fuselage skin, he was able to return safely to base. The feat resulted in the award of a DSO. The dislodged flare that ignited can be seen below the hole in the far side of the fuselage; other flares are in the rack on the right.

 Leonard Geoffrey Cheshire, who was destined to become one of the most eminent pilots to serve with Bomber Command.

The Armstrong-Whitworth Whitley preparing for action - ground staff bomb up at the home station prior to operations over enemy territory.

Stirlings being bombed up. The Stirling was the only one of the three RAF wartime four-engine heavy bombers that was designed as such from the outset. Although it was a good, stable aircraft in flight and surprisingly manoeuvrable, its weight was supported by a wing of short span owing to the design requirement that the aircraft could be housed in the standard pre-war hangars which had 100 feet door openings. The wing area was the chief limiting factor in the Stirling's poor operational altitude when loaded. To shorten take-offs and landings, a very high main undercarriage was fitted to give the wing a sufficient angle of incidence to promote good lift. This made take-offs and landings tricky with the Stirling, which had a tendency to swing violently unless carefully handled.

A Stirling crewmember is dwarfed by the huge bomber. The Stirling was the largest ever bomber at the time of its introduction into service, with a wing-span of 99 feet; a length of 87 feet 3 inches and a height of 22 feet 9 inches.

Richard Rivaz (left) and Leonard Cheshire (right).

Richard Rivaz examining his rear turret guns on the Halifax.

Loading bombs into the bomb bay
of a Halifax.

Pilot Officer Michael Renaut (top) in the cockpit of his Halifax. The wireless operator is sitting in his cubby hole below the pilot and the flight engineer standing behind.

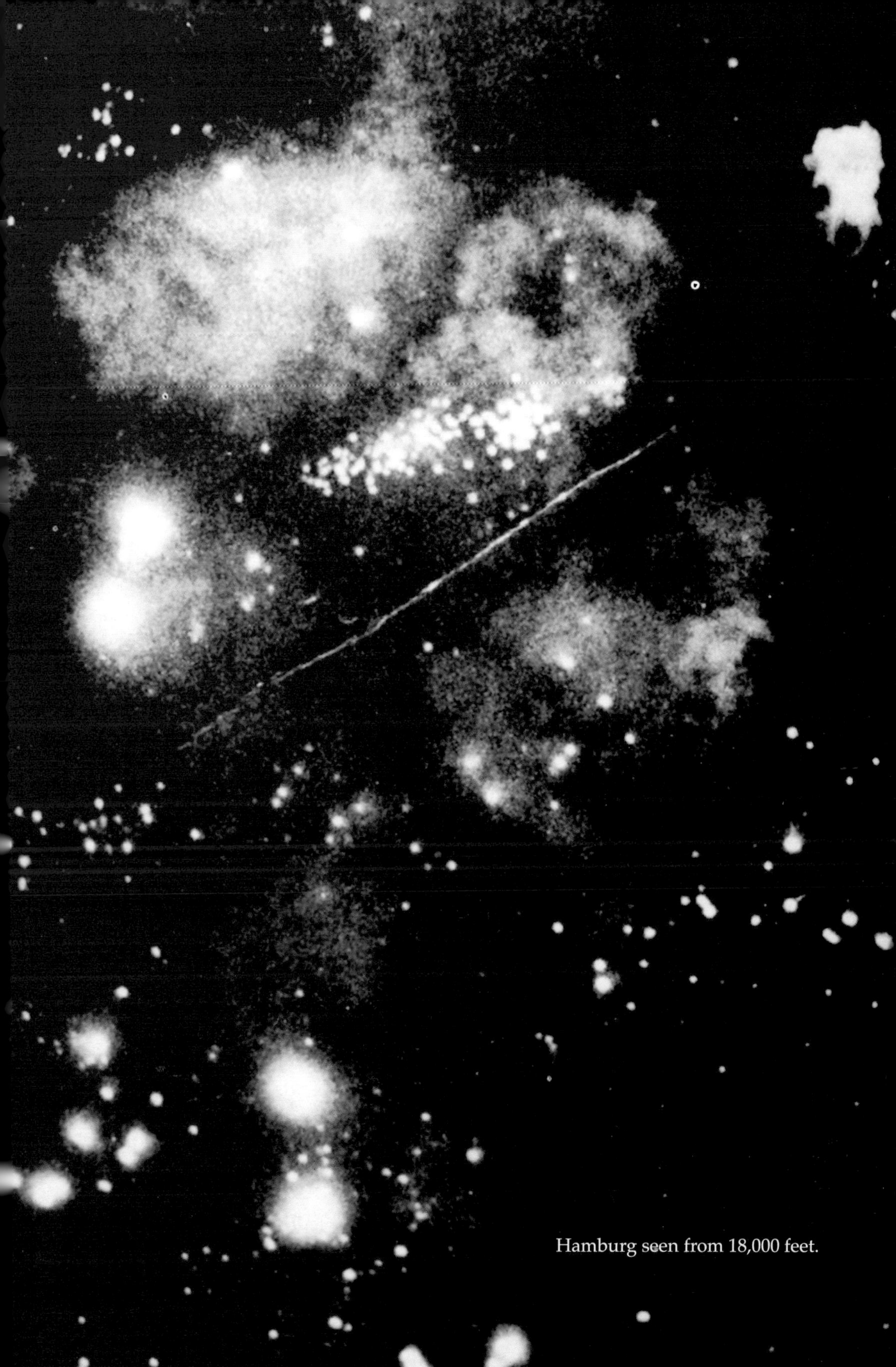

Hamburg seen from 18,000 feet.

Left: Michael Renaut
DFC.

Below: Flight Lieutenant Christopher Cheshire, brother of Leonard Cheshire, on 76 Squadron grins from the cockpit of Halifax B.Mk.I Series II L9530/MP-L at Middleton St. George on 6 August 1941. Painted on the fuselage below is his tongue-in-cheek family crest incorporating various 'Cheshire' elements such as a cheese and grinning cats! L9530 was lost on Berlin on the night of 12/13 August 1941 when it was shot down by flak. All of Cheshire's crew except for the front and rear gunners, bailed out safely and were taken prisoner.

Chapter 8

Busted Flush

I have just been looking over the Manchester. The first twenty can never be used for operations (otherwise than for the purpose of throwing away crews!) and urgent steps must be taken to put right the short-comings. In other words, if the aircraft is required purely for production advertisement it can no doubt remain as it is, but if it is required for operations the following things have got to be done. Parachute Exits. Generally speaking these are not big enough. It is time that the Air Ministry Department concerned realised that in order to make a fully satisfactory exit from an aeroplane it is advisable to take your parachute with you. Bomb-Aimer's View. For some reason best known to themselves they have almost entirely obstructed the Bomb-Aimer's compartment with an enormous cushioned sofa or ottoman. This will be fine if the machine is parked and used for petting parties, but it is quite useless for operations...I think everybody is agreed that the metier of the heavy bomber is night bombing and that any idea of using them by daylight is now a 'busted flush', even amongst those who were not seized of this childishly obvious fact before the war started. It is therefore essential that all heavy bomber armament should be orientated towards night defence ...'

Air Marshal Arthur Harris, writing in September 1940 after visiting Boscombe Down to see one of the Manchester prototypes.

The weather in the winter of 1941/42 was about as unfavourable for bombing as it could be. The records showed that none of the previous fifteen winters had been so bad. The bombers were grounded night after night. When Jack Bushby joined No 4 Manchester Course at 25 OTU, Finningley in September 1941, he was fresh from gunnery school and as yet too inexperienced to realise what he and many others were in for. Born in 1920, his schooldays were filled with romantic pulp magazines and Hollywood air epics of the early 1930s. He wanted to fly and before the outbreak of World War Two while employed in a Fleet Street advertising agency, he became a 'week-end airman'; joining 601 Squadron of the Auxiliary Air Force (AAF) as an Aircraftman 2nd Class (AC2). In 1939 Bushby re-mustered and finally, having volunteered for aircrew, began training 18 months later. No doubt by that time the experiences of 5 Group squadrons already operating Avro Manchesters had been duly passed to higher authority, but nothing was said, which was understandable, since it would hardly have been conducive to high morale to learn that the aircraft on which one was about to commence one's operational career was turning out to be a 'stinker'!

The chief cause of the Manchester's failures lay in its twin Rolls-Royce Vulture X-Inline engines, which at the time of the aircraft's conception in 1936-37 were a new design, yet to be thoroughly developed and tested. The Vulture was, in effect, two sets of cylinder blocks from RR Peregrine engines, mounted on a common crankcase with a 90-degree angle between the blocks, giving it an 'X' cross-sectional appearance. Flight Lieutenant 'Mike' Lewis was one of the six original pilots to fly on 207, the first Manchester squadron. He recalls. 'It was a disaster. The aircraft itself, the airframe, had many shortcomings in the equipment in the beginning but as we found out, Avro were excellent in doing modifications and re-equipping the aeroplane. The engines never were and never did become reliable. They did not give enough power for the aeroplane, so we ended up with two extremely unreliable 1,750hp engines and having to haul a 50,000lb aircraft. We should really have had 2,500hp engines. If you felt that you'd lost one, that was it; you weren't coming home. It didn't matter if you feathered the propeller or not; there was only one way you went and that was down.'

Avro's were well aware of the alarming deficiencies and had immediately began considering a four-engined version, to be titled the Manchester III with four Rolls-Royce Merlin engines, mounted in an extended wing. This re-design was first proposed in September 1939 and on 9 January 1941 the prototype 'Manchester III' - or Lancaster, as it was then named, flew for the first time. Some 7,377 Lancasters would be built so from the Manchester, with all its failures would come the finest heavy bomber to be flown by the RAF during World War Two.

'In the event' says Bushby 'it was almost three months after beginning training before our course really got to grips with the beast. Up until then the OTU training exercises had been carried out on Wellingtons and not until the final phase, in E Flight, did one actually commence real Manchester flying. This took place from Bircotes, a grass airfield near Bawtry, south of Finningley and it was there that I got my first dose-up look, inside and out. As an air gunner, my prime area of interest was the turrets and armament. The rear turret was acceptable and, in some respects, an improvement on that of the Wellington. But that mid-upper turret! There and then I swore they would never get me into it, short of a direct order. The Manchester mid-upper turret was the Frazer-Nash Mk VII, but was popularly referred to as the 'Botha' turret, since its main application up until then had been in that similarly ill-fated aircraft, the Blackburn Botha. Fitted with two .303 inch Brownings, its upper portion above the fuselage was a pointed egg-shape with a flattened rear surface in which latter were incorporated two tiny, hinged, emergency escape doors. Access to this dorsal turret from the fuselage was not easy, to say the least and a hurried exit by this route would have taken so long as not to matter anyway. Hence the above-fuselage doors. I never heard of these ever being used in flight and the only time I tried it on the ground, wearing an Irvin jacket, I stuck fast - and I'm not particularly bulky.'

As on the later Lancaster mid-upper, two arms ending in rollers and linked to the gun elevating mechanism protruded from the outside and followed the contour of an external ramp; the whole arrangement being officially christened GTI - Gun Travel Interrupter - and a necessary precaution to prevent gunners shooting off their own tailplanes and rudders. The wings and forward fuselage were similarly protected by this device.

'Our flying in E Flight, 25 OTU proceeded without incident and we notched up a score or so hours before noticing anything peculiar about Manchesters. Not, that is, until our very last flight on New Year's Day 1942, before moving on to 83 Squadron at Scampton just north of Lincoln and operations the next day. This flight was meant to be an innocuous hour's circuits and bumps, chiefly for the pilot's benefit and utilising dark glasses and a new sodium lamp flare path as a sort of makeshift night-flying-in-daylight experiment. Two student pilots and an experienced second-tour instructor were up at the sharp end, happily carrying out their exercise. We crew members, somewhat superfluous on this local jaunt in daylight, were continuing the previous evening's four-handed cribbage tournament on the main spar casing of L7431.

'The wheels had just been lowered for the umpteenth approach when there came an audible bang - and motley pieces of the port engine began to shoot past our side windows. Up in the cockpit there was a moment of wild confusion, with arms flying out punching feathering buttons in all directions and pilots squirming over and round each other, exchanging seats. Once installed in the left-hand seat the instructor yelled back something about not being able to hold her and to hang on! Even with our minimal load a single engine overshoot proved impossible and, skipping the intervening heart-stopping minutes that flight finished in a crashing, slithering belly-landing across a wet ploughed field; the wild slide being arrested by a sturdy oak tree just as it ran out of impetus. Fortunately no one was injured and considering that with the loss of one engine a Manchester's sole inclination was to adopt the gliding angle of a streamlined brick, the instructor handled the situation extremely well.

'One disturbing characteristic of Vulture engines, apart from their unreliability and failure to deliver the 'urge' was the long, brilliant comet trail of fiery red sparks which they threw out even when running normally. At night these were visible for miles and all kinds of dodges were tried to trap and suppress these trails, without success. No German night-fighter needed radar when Manchesters were about and to my mind this might have accounted for more than a few being easily stalked and shot down.

'About mid-fuselage the Manchester carried a monumental structure looking not unlike a scaled-down Wurlitzer organ and known officially as a triple flare chute. The centre chute was for the high candlepower photo-flash, while the two side chutes were for any other pyrotechnic device needing to be launched. The bottom of each chute was sealed off with a sliding tin cover actuated from the cockpit, which when slid back allowed the flare or flash to drop down and out. Some of our OTU exercises called for the release of an aluminium sea-marker which, when it hit the water, spread a sizeable patch of fine aluminium powder on the sea surface, thus forming a target for gunners to aim at. It was only a question of time before the incidence of Murphy's Law and, sure enough, one day someone let the marker go before the bottom trap was open and it functioned inside the aircraft. On landing, the crew emerged coated in silver from head to foot, while the interior of the Manchester was like a department store Christmas grotto.

'These side chutes were also used for pushing out bundles of propaganda leaflets with which the chairborne Ministry of Propaganda warriors had decided the war would be won. I found out about these the hard way. The first time I filled the chute with bundles of bumf and the trap was operated, the resulting hurricane of slipstream draught back up the chute caused the lot to be 'returned to sender'. The dark interior then became a mass of flapping, whirling paper, plastering itself over me and everything else. As 'Chiefy' remarked caustically after we landed, it looked like 'a blitzed toilet'.

After this incident we just bunged them out still in their bales in the hope that a solid bundle would descend on a Nazi head and do that much good at least. Since a crew member had to devote several minutes to this task, I frequently speculate on how many bombers thus deprived of half of their rearward outlook might have been caught napping by night fighters during such a futile exercise.

'No 83 Squadron began Manchester operations with an easy one to Boulogne on the night of 28 January 1942. Warrant Officer Whitehead's crew prepared to fly our first op on a Manchester. We foregathered after breakfast to see what was ordered for the day. 'Ops tonight. Night flying tests as soon as your aircraft are serviceable. Briefing four o'clock'. This was it! This was the moment the score or so of us who had come from the OTU and scores more like us, no doubt, at other bomber stations had thought about for months. This was the moment we had trained; flown, studied and sweated, 'Ops tonight'. The second-tour veterans were unruffled. They had seen it all before. For we initiates it was best summed up in Dick's (second pilot) sardonic comment, 'Feel like King Canute. Bags of confidence, but I'd like to know just what the hell's going to happen.'

On the short, half-hour test flight before lunch never was an aircraft given such meticulous inspection. I polished and re polished the perspex of the rear turret and spent extra time on the ground cleaning the gun barrels far more than they needed. At lunch we were subdued, each busy with his own thoughts. Afterwards I borrowed a bicycle and pedalled round the perimeter track to where our aircraft stood at dispersal, fitters busy on the wing, their heads buried in the engines and - novel and awe-inspiring sight - a train of linked bomb trolleys stationary beneath the fuselage and the armourers already busy winching up the long, sinister 1,000lb bombs into the belly.

'I went inside and crawled through to the turret looking for something to do, wondering if there was something vital, something important they had told me about in training which I had been stupid enough to forget. The ground crew were good-tempered, although I must have been in their way, elbowing me politely aside and busy with their tasks. Quite honestly I might just as well have been up in the Mess, but something kept me out there on the cold airfield. Subconsciously I felt that close attention to one's equipment before an operation was something one had to do and this really was the only reason for my presence there. Anyway I seemed to have been the only one affected so because none of the rest of my crew put in an appearance. Finally the Sergeant armourer, who had seen many crews come and go and who may well have guessed at what was in my mind, waved at me and I went over and accepted his offer of a cigarette. 'Everything OK?' he asked. It was more of a statement than a question. 'Er, yes, far as I can see.' 'Don't worry, cock. They'll go all right if you need 'em' and he jerked his head at the four guns protruding from the turret. I got the message and, mounting my bicycle, pedalled back to the Mess. I was in time for a cup of tea and joined Whitey (skipper) and Dick, sprawled comfortably in armchairs before the fire. 'Where've you been?' asked Whitey. 'Thought you'd chickened out' said Dick, always ready with a wisecrack. 'Just been down to the aircraft, looking around'. Whitey shot me a keen glance. 'That's the stuff', was all he said but to me it was a medal!

'There was one more thing to be done in preparation. This was a letter to my parents. It was the sort of letter one might expect to be written at the time and under those conditions and as I sealed it, I wondered what to do with it. In the crew room, carefully emptying pockets of every scrap of paper which might be useful to an enemy for identification, I got an idea. Bundled up in heavy flying suit with helmet and mask and carrying a parachute, I wandered into Chiefy's office. 'Chiefy,' I said with what I hoped was nonchalance, 'Forgot to post this and I don't want to trail back to the crew room. Hang on to it for me, will you? I'll collect it in the morning.' 'All right mate,' he said 'They all leave 'em with me.' He pulled open his battle dress blouse and I saw that my letter was only one of several tucked in there. He winked and turned away to answer someone's call. This made me feel a lot better. It was something I had wanted to do, yet felt afraid of embarrassment if anyone noticed it. Here was I, thinking I was the only sentimental fool on the squadron, yet it seemed that many more had the same thought. Later Chiefy told me how many times, after distributing the letters back to their owners next day, he had to keep one or two back and send them on, swearing at the war with each one.

'The greatest surprise about my first briefing was that it was all exactly as I had imagined it from the training films and the cinema. On the wall, at the end of the large room, was a map of Europe; and a ribbon had been pinned with one end located at Scampton and the other somewhere on the French coast. I peered forward with everyone else. 'Boulogne' said someone and there was an audible sigh as fifty men sat back. Whitey was almost smiling. 'Piece of cake' he whispered to Dick out of the corner of his mouth. Then the squadron commander climbed up to the dais and raised his hand for silence. He had it in an instant. 'Well, chaps, your target for tonight is Boulogne.' It flashed across my mind as remarkable that he actually used the exact words just as they did in the films. 'This, as the second-tour bods among you will realise, is an easy one. It is 83's first op with Manchesters and Command just daren't trust you lot any farther than that'. This got him the intended laugh. Then followed target details, photographs of the docks and harbour at Boulogne, anti-aircraft defences - 'Nothing to speak of, chaps. Most of it has been withdrawn lately for the Eastern front'. There was a discussion on weather from the meteorological officer whose appearance was greeted with groans and ironic cheers, which he smilingly acknowledged as part of the ritual pattern.

'Then came what I thought one of the most interesting parts of the briefing and something very few persons apart from those present have ever known about. This was an announcement of what the German Very-Light recognition colours were for that night. I still marvel today that for months on end bomber crews operating over Germany were able to learn the secret German code signals on every night they operated. Somewhere in Germany or occupied Europe, in hourly fear of detection and death, working in some obscure niche where he had access to this vital, secret intelligence, an individual was able to transmit it to England within an hour or two. For a time during 1942 the 'service' ceased to operate and this fact did not go unmarked amongst crews. Ironic and seemingly callous remarks were made quietly at briefing. 'Poor bastard, reckon he's got the chop'. Despite this the 'service' began again later in the year. Whoever he - or maybe she - was - British agent, French sympathiser and resistance worker, or dedicated German anti-Nazi - the grateful thanks of hundreds of ex-Bomber Command aircrews will always be his.

'Last of all came the issue of escape kits. These were small canvas packages containing French money, a map of northern Europe printed on a silk handkerchief, matches, a tiny compass and other small items judged useful for survivors from a shot-down aircraft endeavouring to avoid capture. We had seen escape and evasion training films. We had been given lectures by those who had escaped or evaded capture. It had all been thrilling and mysterious, but now, as I weighed this little packet in my hand, I realised that it was no longer celluloid melodrama or blackboard imagery. This little packet was for me to use in case I was to find myself suddenly alone at night in hostile Europe! It was a sobering thought. I already had a pencil in my pocket whose clip was magnetized and would serve as a rudimentary compass needle. Two of the buttons on my flies of my battle dress trousers were also polarized to act in the same way if torn off and one balanced on the other. This was really spy stuff, exciting, fascinating and not a little terrifying.

'Every detail of that first operation is as clear as if it happened last night. The tension of anticipation, the take-off, the flight over a darkened England, snow-covered fields glistening in the moonlight. Then the silver sheen of the Channel and a few minutes later, over to one side, a darkening border to the sheen that was France and enemy territory. So far we might have been alone in the sky. Not a thing stirred on the ground, not a light shone nor anything to show that this was war and we were penetrating a hostile frontier. I listened to the brief exchanges over the intercom between Whitey, Dick and Bill at his navigation table. Routine passing of terse information.

Checking of course and airspeed. Then, from Whitey, 'Five minutes to go. Think that's it ahead'. Then as I turned the turret to starboard I saw against the black land mass a brief twinkle in the sky. Then another and another. Half a dozen then nothing more. I knew what it was. This was the flak, the anti-aircraft shells bursting against the night sky. So often had I watched it from the ground, but now I saw it from above and aimed, if not at us, then at others of our kind.

'Target in sight, port a bit!' This in an excited crackle over the intercom from Bill. 'OK, I see it. Gunners keep your eyes peeled!' Unconsciously I had twirled the turret to one side, trying to see round the tail of the aircraft. Whitey's sharp reminder jerked me back into a recollection of my job and I slowly rotated the turret from side to side, peering up at the star-sprinkled sky, striving for the first glimpse of a blackness against the black which might be a prowling night fighter.

'Bomb doors open!' The aircraft lurched as the drag of the opening bomb doors came in.

'OK Whitey, starboard now and straight in. I can see the docks'.

'Leave it a minute!' Whitey's voice was terse.

'Then he saw what he had been waiting for. A searchlight stabbed up from the ground, its sword-like brilliance wavering and then settling on one spot in the sky. Instantly another and another lanced up, all converging on the same spot and I caught the dull glint of perspex. Even as we watched, streams of anti-aircraft fire came curling up from the ground to converge on the apex of the searchlight cone. Coloured balls of tracer, at first seeming to rise slowly, hesitatingly and then as they came to our height shooting up to the sky and bursting in a myriad of twinkling reds, blues, greens, blossoming as a lethal night flower. 'Right, they're busy with that poor sod. In we go!' (Whitey's first-tour experience paying off).

The aircraft lurched as Whitey swung it round sharply to port. Below I could see the blackness of the town divided into a dull silver checkerboard of water in the docks area. Bill began his bombing-run monologue, 'Left, left...right...steady at that...right a bit...OK...Bombs gone!' Immediately he said the words the nose went down as Whitey pushed the stick forward. The dive was sudden and vicious and I wondered, in a momentary panic, if we had been hit. Then, as he flattened out, the sky behind me exploded in a twinkling, blazing blossom of flak. I could even see the puffs of black smoke against the stars. 'Flak astern' - this was what I was supposed to report. Whitey's reply had a chuckle in it, 'An' that's the right place for it'. He had known all the time what would happen. Whilst the defences concentrated on the aircraft coned in the searchlights he had begun his run-in. Then, as they switched their attention to us, he had made that sudden, speed-boosting dive and the consequent hail of flak had burst precisely where we would have been if we had carried on flying straight and level after releasing the bombs. 'You cunning old sod', remarked Dick in admiration. 'Got to be in this game, mate...'

'On the flight home I felt dry and drained of energy. I still kept the turret dutifully rotating side to side and assiduously searched the sky, but so tensed had I been on that target run-up that it would have been an effort to have sounded an alarm even if I had seen anything. I realized I was freezing cold and wished we were getting nearer to base. We landed two hours later and as the Manchester swung round on the dispersal point and the engines were switched off I yawned and stretched. All four aircraft returned safely but, after landing, an exclamation from a rigger drew my attention to the central stabiliser of L7423 - a triple-finned Manchester - and in the light of his torch I saw that the whole fabric covering had mysteriously disappeared during flight, leaving only the bare metal skeleton. A few days later an A Flight Manchester, ambling sedately round the Scampton circuit preparatory to landing, nonchalantly shed its port fin and rudder before our astonished gaze. A somewhat asymmetric and decidedly shaky landing was made and the crumpled bits retrieved from a few fields away. And it was on a cold February afternoon that we taxied back to dispersal after a night-flying test and stopped engines. As the ginormous 7 feet 6 inch blades of the port propeller swung to rest there was a musical tinkling sound. A fitter removed the spinner hub and, after investigation, reported that the whole reduction gear had seized and that several propeller retaining bolts had already sheared.

'I went to clear the guns of the live round which was always up the breech of each gun on operational flying. Then it was that I made an awful discovery! In my raw inexperience I had flown the entire operation with the safety catches 'on' all four guns! If we had been attacked it would have been up to me to defend the others and by the time I discovered that the guns were not firing and searched about for a reason, it would have been too late. I felt humbled and bitterly ashamed of my stupidity. Nevertheless it was a lesson and ever after that I always made sure on that point at least. I went farther and every time I flew after that got the skipper's permission to fire a short burst into the air once we were over the sea. Then I knew the guns would be ready to fire at an instant.

'Afterwards came the debriefing. Every crew crowded into the room in station headquarters, in battle dress, roll-neck sweaters and flying boots, dirty around the eyes from helmets and oxygen masks and eyes glinting with excitement. The mug of hot sweet tea handed to each as he entered - Oh, that post-operational tea! Never in my life before or since have I tasted such nectar! Then the call to one of the tables where a frantically scribbling intelligence officer shot his sharp pre-formed questions at us and noted the answers. Did you find the target? How do you know? Could you identify the point your bombs fell? Did you get a photograph? (A swift anxious glance here between Dick and Whitey. The one in his inexperience had forgotten to remind the other who had overlooked it in preoccupation.) We each gave our version of events and probably exaggerated them in the novelty of it all. So much so that on the way to the Mess for bacon and eggs Whitey gave us a little lecture. 'Don't ever tell them anything unless you're absolutely certain', he advised. 'It's dead easy to think you were smack on target and then, when the photograph comes out, find you were miles away. It gets you a reputation as a line-shooter and they just don't believe you after that.'

'It was good advice and we remembered it. He then went on to read the riot act about our having forgotten to release the photo-flash and take the picture. 'Look', he said, 'It's Bill's job to get the picture but he's hellish busy and so am I when the bombs are just away. I don't care if the whole bleeding lot of you scream out a reminder; just so long as we don't forget it again!' We never did, but in doing so we all learned to hate and detest those paralysing few moments flying straight and level over the target after the bombs were released, whilst the million-candlepower photographic flash dropped away beneath and shed its momentary glare over the scene to be recorded on the camera film. Every instinct urged us to dive, twist, duck, weave, anything to avoid the deadly gun batteries below us, predicting our path with menacing accuracy.

'Next night no ops. Down in the pubs and dance halls of Lincoln chests were out a little bit farther and gas masks swung a fraction jauntier. At last we had 'got some in'. Only one, admittedly, but we were the only ones that knew that.

'Almost every target selected during the winter months of early 1942 was a heavily defended one. We went to Essen, Bremen, Cologne, Wilhelmshaven and Hamburg - then started all over again. The second-tour pilots soon came to realise grimly that with anything like a respectable bomb load they just could not attack the 'Happy Valley' (Ruhr) at any more altitude than they had in the obsolescent Hampdens of their first tours. Meanwhile, German flak and night fighter defences had increased in quantity and greatly improved in accuracy; thus at 10 to 12,000 feet we were nicely in the preferred bracket of heavy stuff, yet not out of reach of the light stuff. On one night, Hamburg-bound, we found that our Manchester, R5830, would not even make 10,000 feet without serious overheating of the already suspect engines. To solve this problem two of the six 1,000lb bombs aboard were smartly jettisoned over the Frisians and with a reduced load of 4,000lb a reasonable cylinder- head temperature could be maintained at precisely 11,000 feet. Then there was the occasion when a CSU (Constant Speed Unit) gave up the ghost over Essen. The resulting howl must have terrified the ground defences - it certainly shook us.

'On 12 February came the escape of the *Scharnhorst* and *Gneisenau*, known irreverently as 'Salmon and Glukstein' to all 1942 bomber crews. All day we stood by but were not ordered off. Night fell and we were stood down. The Mess bar had just opened when my crew were called to the crew room by the Tannoy and there told that we had been 'selected' for the dubious honour of flying across the cold, dark, winter North Sea at low level to drop sea mines in the predicted path of the German ships a mile or two off the German coast. It was touch and go whether we would make it in time, but make it we did. History records that at about that time and in that place, one of the ships was so damaged by a mine that she had to put in to port and then remained there for the rest of the war. So perhaps our journey was necessary. On the way back, at only two or three thousand feet over the icy oggin, seven hearts jumped into seven mouths when the second pilot reported the port revolutions counter going off the clock at the wrong end. With computer-like rapidity seven minds had already worked out that with the height loss to be expected immediately as one engine failed and even if she would fly on the other one, we would be several hundred feet below the waves before the situation was stabilised. Fortunately it turned out to be a case of a duff instrument, but never was an engine's note listened to more keenly or with such rapt attention than in Manchester L7464 that night.

'Other crews had their moments too, of course. Although on the face of it the Manchester problem was simply that of an under-powered design suffering from poor performance, peculiar things used to happen to the airframe and equipment. Revs counters would suddenly offer wild and impossible readings, like the night we took off with - according to the panel - 2,8S0rpm on port and 4,500rpm on starboard. Gyro repeaters and artificial horizons, without warning, took off in erratic rotations round the dials; the intercom would become swamped with screaming gremlins and all communications become impossible. Even 'George' (automatic pilot) had occasional fits of gay abandon, trying to perform aerobatics on the way out and back. And there was the 44 Squadron Manchester at Waddington which shed a leading edge on take-off and careered into parked aircraft, somewhat to the detriment of the war effort. With all due respect, later generations of air crews never knew how interesting life could be dodging the Happy Valley flak at eight to nine thousand feet and wondering what was going to fall off next.

'The worst aspect of all this was its effect on morale. No one ever got to the point of actually gibbering with terror every time they saw a Manchester and corny jokes continued to come thick and fast. Yet beneath all this banter was a lack of confidence in one's equipment; bad enough in any sphere but fatal in war operations. There was a continual niggling worry in the back of many minds that sometime soon, maybe, something was going to go wrong at a critical moment, perhaps with fatal results, which must have affected the ability to make instant and correct decisions. Such was the Manchester's evil reputation that it is a matter of record that Flight Lieutenant 'Kipper' Herring of 207 Squadron brought his Manchester (L7432/Z) back from the 'Big City' (Berlin) on one engine, below 1,000 feet all the way and was awarded an immediate DSO. Unanimous 5 Group opinion was that a decoration was never more richly deserved. Rumour had it that at one point Herring's crew had nothing left to throw overboard but their trousers and were starting to dismantle the airframe to save an extra pound or two of weight to scrape home across the North Sea.[58]

Harold Southgate, a Manchester pilot on 50 Squadron, recalled: 'Having raided Rostock on 24 April 1942 with comparative comfort and with very little opposition, another raid two nights later was regarded as 'a piece of cake'. Although a long trip of some hours and on a more or less direct route, the flight to the target was uneventful. The Manchester was loaded with one 4,000lb bomb plus several canisters of incendiaries and feeling - stupidly - rather brave, I decided to bomb from 4,000 feet, the minimum altitude from which a 'cookie' could be dropped safely. Having started the bombing run, all hell was suddenly let loose from below with flak bursting all around us and shrapnel hitting the aircraft. In an effort to avoid this surprise reception the bombing run was continued in a dive - contrary to Bomber Command's laid down procedure - and the load was eventually released from about 2,000 feet.

'Our Manchester was nearly blown out of the sky by the detonation of our own bomb plus sundry others which had been released correctly by other aircraft from far above. We took hits in several places and the controls were becoming unmanageable, particularly the rudder. Fortunately, on board for a familiarization trip was a young flight sergeant pilot. He really had to earn his keep on the way home, as both of us had to push hard on the rudder controls to keep the Manchester on a straight course. This was a really hard struggle for more than three hours. Approaching our base we received a priority for landing and with a great deal of effort got the Manchester down safely. It taught me two lessons: always obey orders and never underestimate the Germans - who had completely reorganized the defence of Rostock in two days.' [59]

An attack on a second Baltic port - that of Rostock - on four nights running from 23 April to 26 April was not only, as at Lübeck, to make a port which was used as a supply base for the Russian front unusable, but also to put out of action the most important of the ring of Heinkel aircraft factories in and around the town. These factories had been built not long before the war; one of these factories was the only place in northern Germany where Heinkel aircraft could be assembled. It was Heinkel who in recent times brought prosperity to Rostock. The influx of skilled workers into his factories brought the population of the town up from 89,000 in 1933 to about 115,000 in 1939. More workers certainly came to the town after the war began. Flight Lieutenant A. F. Taylor DFC set down his thoughts in *Flight to Rostock* [60]:

'This raid'll be a good show,' I said, calling up Jimmy on the intercom. 'The visibility is wizard and we can go in as low as you like.'

'We're about five minutes late,' he said, 'the attack should start about a quarter of an hour before we get there. I'm coming forward in a few minutes.'

'I heard the door open and turned to see Jimmy at my side. We grinned at each other and I pointed ahead where light flak had just started to weave angry streaks in the sky. A few minutes later incendiaries began to go off and little flickers of flame marked the beginning of fires. As we neared the target the tempo of the attack increased. Two pale searchlights wandered inconsequently here and there. The flickers of flame had become ugly red fires, merging at places into a great molten blaze. As we crossed the coast the whole target was lit by fires which reflected a ruddy glow in the water of the harbour. High-explosive bombs were bursting with great yellow flashes. A burst of light flak was fired at us, little red and green balls drifting up slowly, lazily, at first and then ripping past our wings in vivid streaks. Several black puffs about a hundred feet to port marked an ineffectual salvo of heavy flak.

'This is a piece of cake,' said Jimmy from the bomb-aimer's position. 'Bomb doors open.'

'Bomb doors open.'

'Left left, steady, left left, steady, steady now. Bombs gone.' 'Flash gone,' said the second pilot from the flare chute immediately afterwards.

'I held her steady until I saw the flash go off in a blinding white glare below, then closed the bomb doors and banked over to have a look. Among the criss-cross network of the streets a great wide path of little white scintillating lights marked our incendiary stick. As I flew round, watching, some of them glowed red and little wicked tongues of flame spurted out. The whole town was blazing now, the giant fires merging into one huge red inferno. For many miles we could see the glow in the sky that marked Rostock. It was just as easy as that. Crossing the coast again we began a climb back to oxygen height and Jimmy retired regretfully to his 'office,' with its multitude of maps, instruments and pencils.

'After a while I pulled the revs, down and began to lose height slowly. The second pilot pumped some oil and came forward to take over. A gentle swaying to either side of our course told me that Bob was keeping his rear turret moving. I went forward to the bomb-aimer's position and gazed vaguely at the sea. I was very tired and the prospect of coffee, bacon and eggs in the mess seemed beautifully welcome.

'The fighter attacked with terrifying suddenness. An awful grinding, splintering roar accompanied by a blaze of light as the tail of our aircraft caught fire. The whole aircraft shuddered and for a moment I thought that both motors had seized up simultaneously. I felt a searing pain across my right thigh. An incendiary bullet crashed through the floor at my feet and whizzed round like a squib. I jumped backwards to the step by the second pilot, jerking my inter-com plug out as I went.

'I don't suppose it took two seconds for the second pilot and me to swap over. My memory of the next few seconds is blurred. I must have acted quite mechanically, as my mind was dazed with shock. The whole aircraft, wings, engines, cockpit, everything, was lit brightly by the light of the fire at the back. The two motors were running smoothly. We were in a steepish dive and the control column was useless - the connecting-rods were shot away. I wound the tail trimmer back and put up the revs on both motors. The nose rose sluggishly and we levelled out as I juggled with the trimmer. Just then bullets began hitting the aircraft again, with that dreadful staccato tearing noise. I trod hard on the right rudder and as suddenly the sound ceased. Great red, angry streams of tracer bullets tore under the port wing and disappeared into the night.

'The second pilot was standing calmly beside me. They were all calm, the whole crew. No doubt they were as frightened as I was, but there was no panic. They all carried on with their jobs coolly and efficiently. I pulled the second pilot towards me by his helmet and shouted in his ear. 'See if anyone else is hurt; get the front gunner out of his turret and put that fire out.' He nodded and carried on. The airspeed indicator was registering nothing, the artificial horizon flapped stupidly at a crazy angle and the gyro directional indicator was spinning wildly. Fortunately there was little haze and sea and sky made a vague horizon. I picked out a star dead ahead and flew as steadily as the wallowing aircraft would permit. The compass showed we were about 300 off course, so I turned the aircraft awkwardly and fixed the star in its new position. The aircraft was flying very sluggishly; the only way to control the fore and aft attitude was by manipulation of the engines and the tail trimmer. To keep laterally level the wheel had to be held three-quarters of the way to port. The aileron trimmer was not working and the port landing-wheel hung drunkenly out of its nacelle. By way of experiment I tried the flaps: there was no response.

'The fire at the back was still blazing furiously. My right thigh, which had not worried me since that first searing pain, was now aching. I put my hand down to rub it gently and when I looked at my hand afterwards it was covered with blood. The front gunner was now out of his turret and standing beside me.

'Jimmy's used up the fire extinguisher and the fire's not so bad now, but too hot to get anywhere near. We've taken out the astro hatch and the draught may damp it down.'

'I learned from Jimmy that we were about an hour's flying from the English coast. The longest hour I've ever known. In continual fear of further attacks from fighters we struggled lamely to maintain height and course. The wireless operator had got the wireless going, bashed out an SOS and maintained touch with base. Jimmy, working by the light of the fire, gave occasional alterations of course. The second pilot and front gunner did what they could about the blaze at the back. This gradually died down till, when we landed, it was just smouldering. The gaunt blackened framework was left open to the night; in many places, especially round the rear turret, the geodetics were melted clean through. How the turret stayed on was a miracle of sound construction and careful maintenance. Good maintenance too, kept the two motors running smoothly under the increased strain. After a little more than Jimmy's scheduled hour the English coast loomed up. My leg was feeling far too stiff and painful to contemplate bailing out, so I determined to land at base. The petrol gauges were not working, but as we had only turned on the nacelle tanks an hour previously I was certain we had enough to get back.

'We altered course at the English coast and started on the last lame leg for home. The crew refused my offer to let them bale out near the airfield, so win or lose we were all in it together. By the time we reached our base the sky was just turning grey with the first light of dawn, but on the ground it was still quite dark. The R/T was useless and as no lights worked we were unable to flash our identity letter in a request to land. We were some time overdue, anyway, so the chances were that everyone else was down. The only thing to do was to go ahead and hope. I have no clear memory of how I got that aircraft down. I re¬member telling the crew to wedge themselves in tight for the crash and I remember doing a wide, awkward circuit, fearfully losing height - knowing that there could be no question of going round again if I misjudged it. The airspeed indicator was not working so I had to judge the speed as roughly as I could from the aircraft's altitude, keeping it rather nose down for fear of stalling. As we crossed the boundary hedge, only a few feet up, I pulled up both main petrol cocks and cut the motors. The nose fell sharply and we hit the ground with a rending crash. Dust and earth flew into the cockpit and half choked me as we slithered about fifty yards, slewed sideways and stopped. Then everything was still and quiet - we were safely down.'

On 29 April 1942 Manchester VN-N, L7516 of 50 Squadron took off from Skellingthorpe on a mine laying operation in Kiel Bay. Flight Sergeant Hector J. MacDonald was the WOp/AG.

'All went well until we turned for home, when it was discovered we had a hang-up with one mine. Simultaneously we were attacked by a night fighter. Our three gunners, front, rear and mid-upper, returned the fire but we were severely damaged with the port engine ablaze and the landing lights full on, obviously a 'short' as a result of the fighter's fire. We were at about 10,000 feet but the skipper, Flight Sergeant Tim Willett DFM skilfully nursed the aircraft down and we stood by for ditching. Unlike the ditching we expected, we came to a sudden stop. We had hit a sandbank. Sergeant Cecil John Scott and Sergeant David Alexander Williams RAAF, the rear and front gunners, had been killed. The rest of us in various states of damage scrambled out and expected to see the dinghy but it did not appear - another casualty. The Manchester was well ablaze by now; the ammo was popping off and oxygen bottles exploding, so we decided to get away relying on the Mae Wests. We knew we were not far from a coast as the horizontal beams of searchlights picked us out and, after about two hours, a motorised rubber boat appeared, hauled us aboard and back to shore which turned out to be Sylt.

'Flight Sergeant S. E. Packard, the second pilot and me were taken to the Luftwaffe hospital at Westerland, eating en route the unappetising colours of the day. After being operated upon, we were told by the German staff that the aircraft that had attacked us was an Me 110 and that the rear gunner, Unteroffizier Schubert, had been so severely wounded in the stomach that he had died in the same hospital. Schubert was shot in the abdomen and their aircraft was so badly shot up by us that it went down but managed a crash landing which aggravated Schubert's wounds to a fatal degree. It then became clear to me that not only had he shot us down, but we had returned the compliment. We also had a visit from Oberleutnant Günter Köberich, Staffelkapitän, 11./NJG3, the pilot of the Me 110 and, although we were at a disadvantage language-wise, he was very friendly. He was accompanied by another pilot who had himself been shot down by a Wellington a few weeks earlier. Köberich continued to increase his score until Easter Sunday 1944. He was serving at the airfield at Quakenbrück when, around lunchtime, the American Air Force came over carpet bombing. The ceiling collapsed just over where Köberich was standing and his was the only death in the cellar.

'After spending fifteen months in various PoW hospitals, I eventually met up with the remainder of the crew in Stalag Luft VI and from them I learned of the events after I left them in Sylt. As the Manchester was only partly submerged, up to the underside of the wings, a salvage party had been sent out and they discovered the hung up mine. This was one of the anti-sweep type and, in attempting to recover it; it blew up and killed all the party.'[61]

In May 1942 Jack Bushby was detached on a gunnery leader's course, 'having survived - and I select that word deliberately - ten operational trips on Manchesters, with a total of over 100 hours on the type. It was with a feeling of heartfelt relief that I returned at the end of the month to find them gone, replaced by Lancasters, although it was saddening to know that many of them had gone down into the night over Germany, taking with them more than half of the crews alongside whom I had trained at Finningley.'

On the night of 3/4 June when 170 bombers were despatched on the first large raid to Bremen since October 1941, eleven aircraft were lost. They included a Manchester on 50 Squadron flown by Flying Officer John Frankland Heaton, which fell to a night fighter and crashed near Apeldoorn. Sergeant Ken Gaulton, the wireless operator, recalls:

'Our aircraft took off from Swinderby at about 2200 hours on 3 June. We flew across northern Holland and dropped our bomb load on Bremen. I switched to the aircraft inter communication system to advise the pilot that we were cleared to return to our base, this information having been received on the 0230 hours' broadcast from Group HQ.

'On our return flight we were attacked by a Messerschmitt 110. The starboard wing of our aircraft was burning and the pilot advised that he was going to dive in an attempt to 'blow out' the fire. This did not succeed. The German aircraft did a victory roll near the tail of our aircraft and Sergeant P Buttigieg, our tail gunner, shot him down. I was amazed to hear the yelling from the tail gunner who was engaging the Me 110 with his guns.

'After diving for thousands of feet, I requested the pilot's permission to have the tail gunner and mid-upper gunner join me to prepare the rear escape hatch for evacuation. This was done and we jumped in turn; firstly the tail gunner (the only married man in the crew) then the mid-upper gunner and then came my turn. The aircraft kept on diving and crashed, killing Heaton, Pilot Officer John Ross Steen, the co-pilot, Pilot Officer Harold William Sheen, the navigator and Sergeant S Thomas, the front gunner; all of whom were in the front of the aircraft. I left the aircraft when it was slightly under 1,000 feet, quickly pulled the ripcord and was promptly knocked out by the chest parachute striking me under the jaw. I landed in the Zuider Zee on an ebbing tide and speared up to my chest in mud. An Alsatian dog woke me by licking my face and its owner took me to a medical doctor at about 5.30 am. I was unable to walk. The doctor quickly established that the man was a collaborator and was therefore unable to hide me. I was transported by car to Arnhem where I was interrogated by the Gestapo and then by train to Amsterdam, where I was gaoled in the Amsterdam watchtower for four days. While I was there a German captain from the fighter squadron visited me about 6 June and advised me that our aircraft had crashed on to a hunting lodge near Apeldoorn (a lodge owned by the Dutch Royal family). He told me that we had shot down one of his aircraft, killing two airmen. He claimed the Germans were two up as four of our crew had been killed.'

Another two 50 Squadron Manchesters were lost on the night of 6/7 June. Sergeant Leonard Thomas Baker's aircraft was shot down by Ludwig Becker at 0044 hours and crashed into the North Sea off Ameland. (Hauptmann Becker was killed a year later, on 26 February 1943, having destroyed 44 aircraft). Pilot Officer A. D. 'Don' Beatty and his all RAAF crew ditched their Manchester off the coast of the Dutch Friesians after their Manchester developed engine problems. Altogether, nine aircraft - three Manchesters, three Wellingtons, two Stirlings and one Halifax - were lost from the 233 aircraft despatched. German night fighters shot down both Stirlings. W7471 of 7 Squadron was downed between Blija and Holwerd and N3761 BU-E of 214 Squadron, flown by Flight Lieutenant Reginald William Arthur Turtle DFC, whose crew were on only their second operation, crashed in the North Sea off Terschelling. All the crew was killed.

Endnotes Chapter 8

58   Quoted in *Guns In The Sky; The Air Gunners of World War Two* by Chaz Bowyer (J. M. Dent & Sons Ltd 1979). Manchester L7432/Z was lost with Flying Officer John F. Heaton and crew on 50 Squadron at Skellingthorpe on the operation on Bremen on the night of 3/4 June 1942. The aircraft crashed in flames at Beekbergen near Apeldoorn at 0233 hours. Heaton and three of the crew were killed. Sergeant Peter Buttigieg the tail gunner, a Maltese national, Sergeant Ken Gaulton and Flight Sergeant John Farquhar bailed out and were taken into captivity. Buttigieg had tried to get back to England but was betrayed and fell into the clutches of the Gestapo. He survived the war. See *Bomber Command: Reflections of War - Cover of Darkness 1939-May 1942* by Martin W. Bowman (Pen & Sword 2011).
59   Quoted in *Experiences of War: The British Airman* by Roger Freeman (Arms & Armour 1989).
60   Quoted in *Slipstream: A Royal Air Force Anthology* (Eyre and Spottiswoode, London, 1946).
61   As a result of my wounds I was repatriated in October 1943 on an exchange in Gothenburg. *Intercom; the Official Quarterly Magazine of the Aircrew Association,* autumn 1992.

Chapter 9

The Tail Gunner's Story

My brief sweet life is over,
My eyes no longer see,
No summer walks,
No Christmas trees,
No pretty girls for me,
I've got the chop, I've had it
My nightly ops are done,
Yet in a hundred years from now,
I'll still be twenty-one.

'Requiem for a Rear Gunner', **Sergeant Ralph Wilson Gilbert, air gunner on 158 Squadron from his book of the same title.**

It was Cologne again on the night of 1/2 March 1941 when 131 Blenheims, Hampdens, Wellingtons and Whitleys were detailed to bomb two targets. Richard C. Rivaz on 102 Squadron could not say if he was altogether looking forward to this particular target, as the last time he went there, on the night of 12/13 November 1940 - incidentally, the last trip he had done - he and the rest of Leonard Cheshire's crew had the incident of the exploding flare. Even so, Rivaz was longing for another trip. On this March night he was to fly with a new captain, 26-year old Squadron Leader Clive Eugene Erwin Florigny, who was his flight commander - 'a very decent chap' and whose pet remark was 'God bless my soul!' which always came out when he wished to show astonishment'. Florigny's younger brother, Alan Addison, a pilot officer on 10 Squadron at Leeming would also be flying this night, at the controls of a Whitley V.

'I had been acting as adjutant for the last few days, as our adjutant was on leave' recalled Rivaz. 'I hated the job, as it meant being tied to the office with no chance of flying and anyhow I knew very little about it. We knew we were on by ten o'clock and had plenty of time to do our DIs and NFT in the morning. We landed from our NFT by twelve o'clock and were free until briefing, which was at three o'clock. I am always restless before a trip and can't settle down to anything and find it difficult to concentrate: if I play a game of tennis or squash I find it very difficult to keep my mind on the game; I feel as I do before going to the dentist or before the beginning of a match.

'The group gunner leader rang up during the morning to ask if he could fly in 'S' which was our aeroplane. 'S' had been given to the squadron by Ceylon and as he had been in some journalistic job there before the war he wanted to fly in her and write an article on the trip for Ceylon. I was delighted when the Squadron Leader said I was flying with him and that he would not change his gunner: although, as I said, I was not particularly looking forward to the trip, I would not have missed it for anything.

'There is a thrill and pride of achievement about operational flying that I can't describe. I feel at the end of a trip that I have really done something; that I have accomplished something great. I feel that I have defeated something, too. I don't know what it is; whether it is my own fear or death or what but I always feel at the end of a trip a wonderful sense of calm and peace. I have always loved England, but never as much as I do when I return from a trip over enemy territory. There is always the feeling before setting out that it might be the last trip; that one might be blown to pieces or have to bailout over there. I don't think I am afraid of dying, but I am afraid of being taken prisoner and I certainly have a terror of being badly wounded but, worst of all, I am afraid of being captured. It is the uncertainty of it all; not knowing how much longer I should have to be there and wondering all the time what would be happening at home and who would be there when I got back.

'I hate to see caged animals or birds although we are told they are happy as they have nothing to fear and have regular food. But surely fear with an animal is an instinct without which it would have no self-preservation. Fear with us is not an instinct but something very real and something never to be forgotten. Some people have never known fear, except possibly by what they call 'a fright'. Real fear hurts: it possesses your whole being: it grips you and makes you conscious only of yourself: it makes you lose count of time! The coward in me sometimes says 'Well, if you are taken prisoner you will have nothing more to fear. You will be safe, even if you are not at home.' But no! I would rather fear and live my own life and be surrounded by all the risks and uncertainties of war.

'Briefing was longer than usual, as some of the crews had not been to Cologne before. It was to be a big effort [131 Blenheims, Hampdens, Wellingtons and Whitleys to two targets in Cologne] and our target was on the east of the Rhine: there were enlargements of our target maps, chalked in colours, on the blackboard. I studied these very carefully, noticing particularly the shape the river made as it curved through the town. I wanted to have a mental picture of this map, as sometimes you get a view from the tail that the others have not seen. The Met report was not too good: the winds were high and there was the possibility of a front before reaching the target. However, the Met man held out good hopes of Cologne itself being free from cloud.

'After briefing, we had tea, with a couple of boiled eggs. I was not feeling hungry, but I ate all I could; partly to keep warm and partly because I did not know where or when the next meal would be forthcoming. After tea, I went to my room to put on some warm clothes. I always wear as many clothes as I can: there is no excuse for feeling really cold in the air provided plenty of clothes are worn. It is often impossible to keep warm but there is a great difference between feeling cold and feeling unbearably cold. The nose and tail turrets are the coldest positions in an aeroplane and gunners usually wear more clothes than the remainder of the crew. I wore vest, pants, shirt without collar or tie - the latter a precaution in case we came down in the sea and a collar is apt to shrink and strangle one - three pullovers, a roll-top sweater and four pairs of socks in addition to tunic, flying clothing and scarf.

'Before leaving my room I had a final check up to see that I had not forgotten anything. Pockets empty of all papers. (This, by the way, is an order, not merely a precaution and is intended in case one is taken prisoner); revolver; torch; pipe; tobacco pouch full; thermos flasks of hot coffee; extra scarf and gloves; money; clasp-knife; matches; these are the essentials I always carry. As I filled my pockets I tried to keep my thoughts steady: a lot of imagination can be a curse. My mind had to be free and clear for the next few hours. It would have to be quick and alert. All my movements while I was getting ready were intentionally unhurried: I had plenty of time and there was no need to hurry. I did not want to feel rushed and flustered: I knew that I had to sit still for several hours and a last-minute rush would not have helped. We were due to takeoff at 18.50 and at about five o'clock I set off leisurely towards the crew room, where other air crews were arriving with their kit. The atmosphere in the crew room before an operational trip is always exciting and tense. Everyone is jovial and friendly and seeming to give of his best but all the same there is a feeling of strain and one wonders how much people are acting. However we are on the job now and personal feelings do not matter.

'While we were getting into our flying kit, the CO came in with the Station Commander. They, too, were more friendly than usual and stayed talking to us until we were ready to move out to the aircraft. We had another five minutes before we needed to move, so we gathered into our respective crews and talked over again the plans of flight.

'As I said, I was flying with the flight commander. He was a very experienced pilot, although this was only his fourth or fifth trip. He was fairly new to the squadron, so I did not know him very well but what I had seen and knew, I liked. He was about twenty-three, but looked older: he had black hair and black moustache and large dark eyes. Our wireless operator, Arthur Bush, had actually finished his tour of trips but had asked particularly to fly on this one as 'S' was his old aeroplane. His pilot had been posted and he was waiting to be sent on a rest. He was nineteen, tall and very cheery: he gave the impression of being casual and even lazy but when there was a job of work to be done that concerned him he was always there. He was an expedient wireless operator and I knew he could be relied upon in an emergency. Our navigator, Bill Hood, was well on the way towards the end of his tour; he was, short and fair and always grinning; he, too, was thoroughly reliable. The second pilot; Sergeant Martin, was new to the squadron; this was to be his first trip. He was rather shy and quiet.

'It was time to move out to the aircraft. The transport was waiting outside the crew room and we piled in with all our belongings: parachute packs, maps, Very cartridges, thermos flasks and so on. I was beginning to sweat under all my layers of clothing, in spite of my leisurely movements. The ground crew were waiting for us by the aeroplane and helped us stow our kit.

'It was still light, so we could see comfortably what we were doing. I gave a final polish to the perspex round my turret before climbing in. I sweated even more getting into the turret, which was a hell of a job, clothed as I was. However, I was in at last; and had a final check up inside. Everything was as I had left it in the morning and seemed OK. My parachute and flask of coffee were stowed in the fuselage, just outside the turret: I could reach both by opening the turret door at my back and stretching behind me. Inside, I had an extra scarf and gloves, chewing gum, chocolate, barley sugar, biscuits and clean rag. I put on my helmet and plugged in the intercom. The engines were being run up and tested: the whole tail shook and vibrated and the ammunition rattled in its tanks. The ground crew were standing by, watching: one stood too near the slipstream and had his hat blown off. It was rolling over and over behind the aeroplane and he was chasing it. He, too, was being blown along by the slipstream and when he had retrieved his hat he had to struggle to one side, leaning forward until he was clear. A large pool of water by my turret was being thrown up into a fine spray and some bits of oily rag were flying about in the air. I noticed all these things in detail as each engine was being run up separately.

'My mind was absolutely clear. I no longer felt an individual but somehow part of a scheme working almost automatically. All day I had been preparing for this moment and now I was ready. I could control my actions, but I could no longer control what happened to me. I must sit and wait. I was in the hands of my pilot and God and must sit and wait. I don't think pilots realize or appreciate the trust their crews have to put in them. Four or more men are literally under the control of one man: upon his quick actions and thinking depends largely their safety. If a pilot says 'Jump,' the crew cannot say 'Is it necessary yet?' but must jump. It is not so bad when you know nothing about flying but when you have flown and understand the principles of flying, you find yourself at times apt to criticize the actions of your pilot. This, I know, is wrong as he obviously knows more about his part of the job than the remainder of the crew, but nevertheless it is so. A good captain of an aeroplane will not criticize the actions of his crew unless he knows them to be wrong. He will check up, yes; it is his job to know what each member of his crew is doing - the captain of an aeroplane is as much in command of his aircraft as a naval commander is in command of his ship.

'Many people cannot understand how a sergeant can be in command of officers in his crew. Rank in the air, in a sense, does not count: the sergeant as captain of an aircraft is captain because of his experience and skill as a pilot. It has been decreed and rightly so, that the captain of an aeroplane should be the pilot and not the navigator, wireless operator, or air gunner. It is obvious then, that as each squadron has a certain number of NCO pilots, a captain will often be of non-commissioned rank and may sometimes have one or more officers in his crew. Some of the best captains I have known have been NCOs.

'We were now ready to move off. The squadron leader called through the intercom and asked if I was OK and ready to move. I heard him call up each member in turn. I liked the sound of his voice: it was clear and calm and confident. Again, I often wonder if pilots realize how much the sound of their voices down the intercom can mean to their crews particularly that first call-up. Some pilots do it merely to check the intercom; others to make certain that each member of the crew is in his position but others and these are the ones I like, call through in such a way as to make you feel personal and of some importance: they make you feel that your comfort and feelings matter and are of importance to them. The squadron leader gave this impression and I liked him for it.

'He turned the aeroplane on to the perimeter track. The ground crew ran round out of the way of the slipstream and watched anxiously and with pride. There were fitters, riggers, electricians and armourers there. They too had been working hard all day preparing for now: they had done their work and done it well. Each man there had a responsible job and knew that the safety of the aeroplane and the lives of the crew depended to a great extent on how he did his work. They had taken a pride in it and knew it was OK. They were free now, but were in no hurry to go; they would watch us off the ground and out of sight before they went to their tea. As we moved away they gave us 'Thumbs up' and waved. I moved my guns up and down in reply: they were beautifully smooth and I knew I could rely on them. As we moved round the track we passed airmen who stepped and waved. Some were on bicycles and got off. Many of them were probably envying us and wishing they had the opportunity to share our experiences.

'When we reached the take-off point we had to wait our turn and others were lining up behind us; it was beginning to get dark and I could not see who they were. I again checked over my turret. Guns leaded; sight OK; leading toddle in stowage position; spare sight bulbs; yes; everything was as it should be. We moved slowly round ready for take-off. I could just see a group of people watching us: there was the station commander, CO, padre, duty pilot and several others. They would stand there until the last plane was off and then move off to their several jobs or relaxations.

'The engine reared as the pilot opened the throttles with the brakes on; I could feel the tail beginning to lift as the aircraft strained against the brakes. We moved slowly at first and then very rapidly. The tail was well up and there was a swaying movement as the pilot kept her straight with his rudders. The ground streaked past and seemed to be dropping; the watchers disappeared into the dusk below. We passed over the edge of the aerodrome and I could see the lights glowing red and yellow. As we climbed, the ground appeared darker and the colours faded. The light up where we were was brighter, which, by contrast, made the ground appear darker than it really was but my job was not to watch the ground: enemy fighters have been known to lurk ever the aerodrome and wait for the unwary crew.

'A few months previous to this we lost an aeroplane and crew this way. They had taken off and climbed to about five hundred feet, when a Jerry fighter attacked them from the beam. The first the crew knew was that tracer bullets were hitting them in the fuselage and wing. One engine was hit and caught fire and the aeroplane crashed in flames about a mile from the aerodrome. The only man saved was the captain, a boy of twenty. I met him a year later at a dance and his nerve was still gone; although he was flying as an instructor he told me he did not feel equal to flying on operations again for some time. It is pitiful to see a boy's nerve broken. Anyone whose nerve has gone goes through hell at times and the cause of their breaking comes before them more vividly than at the actual time. The shock at the time of an accident is usually felt afterwards and is not necessarily noticed at the time but at the recurring memory when the imagination has had free time to work, the mind goes through all the mental torture that the imagination can conjure. Physical pain is quickly forgotten; but mental anguish brews and returns until it is almost stifling. A boy's mind should be free and gay and should not know these horrors but the mind of a boy who has tasted war is no longer young, but has outgrown his body.

'We were still circling the aerodrome and climbing and it was getting lighter instead of darker the higher we climbed. The ground appeared as a sort of grey-green colour and seemed very remote and unreal. The aerodrome beacon was flashing red and I made a mental note of the letters -'A.R.' I knew we were still over the aerodrome as I could see the perimeter and obstruction lights below, but that was the only indication of its whereabouts. Hangars and buildings had merged into the ground and were invisible.

'It seemed strange to think that life down there would be going on just the same. In the watch office they would have chalked up our time of take-off and would be awaiting news of us on our return journey. In the Mess, people would be drinking their cans of beer and talking and laughing and listening to the wireless and some would be playing billiards or snooker, or table tennis. There would surely be a pair on the squash court, as this was the popular time for playing and others would be going to York or Harrogate or Ripon for the evening. All these things seemed very far away.

'Up here we were five men working together and for each other and we were all working for the same purpose: to reach Cologne, identify and bomb our target - doing as much damage as possible - and reach home safely. Our lives depended on each other and each one of us was indispensable. The captain's job was to fly the aeroplane to Cologne and back and make a safe landing, avoiding flak and searchlights as best he could. The navigator's job was to tell him what course to fly and to do this he had to calculate and check. Winds are his chief problem: he has to know the exact strength and direction of the wind at whatever direction he is flying. The wireless operator's job is to keep a listening watch at his set: he may be required to send out messages, or to help the navigator by getting loop bearings or fixes but the majority of his time is spent listening. The second pilot sits next to the captain and usually just sits!

'If the visibility is good he will probably map-read. He may fly part of the way. He is there to learn and will be watching and noting all that goes on. My job is to keep a continual look-out for enemy fighters; to report on the movements of searchlights and positions of flak bursts; and to act generally as the driving mirror for the pilot.

'Just turning on the course new, navigator' I heard the captain say this down the intercom and could see the lights of the aerodrome dead below us.

We were on our way new and were still climbing. The sky above us was a green-blue and the western sky was lit by a glorious red sunset. The red glow tinted the edge of my gun barrels and the Perspex round my turret a bright red colour.

'I was thrilled with the beauty and called through to the captain telling him about it and asking him if he could see it. He replied that he could just see the edge of it. They would have lost the sunset from the ground by now but up here it was as vivid as the ground was obscure. On the ground one is not always conscious of the transition of light to darkness unless one's attention is attracted to it by some phenomena such as a bright sunset, or unless one is trying to race the light in a car. But in the air one is in the change; it is all around one. There are no shadows or trees, or hills, or houses to hasten the darkness: the light is about one and seems to linger as it changes colour across the sky; dawdling its way to the west and sinking lower and lower until it shows only as a faint glow or streak on the horizon. I had to keep my eyes moving: it does not do to have one's gaze fixed too long in one spot, as one is apt to get dazed and anyway, the rear gunner's job is to search the sky and not gape at sunsets. The stars were beginning to show as the sky grew darker. It was impossible to see now where the sky ended and the ground began. The sunset was fading and there was just a yellowish glow left above the bank of clouds which a few minutes before had seemed alive and glowing and were now a hazy purple-grey bank in the distance. It was as though our last link with the ground had disappeared: we seemed more alone than ever.

'There was silence down the intercom; there was nothing to say. Our work was only just beginning and we had plenty to think about without having to speak and anyway, there was nothing to say. The sweat on my body was cold and uncomfortable, but I should feel more comfortable when it was dry. I took off my gloves and unwrapped a piece of chewing gum to chew and wished the peppermint taste would last longer! I did not keep my gloves off for long, as it was beginning to get cold. I called through to the captain and asked the temperature - minus 10°. This by the way was Centigrade. Yes, it was going to be cold tonight; devilish cold. We should soon be needing oxygen.

'The second pilot said he could see the coast line ahead and a few minutes later he said he could definitely pin-point us. The captain and navigator had a natter as to whether they should alter course now or wait a bit! They decided to wait. Bill said he thought we might be a bit out to the left. Good old Bill! He certainly knew his job. He had got the winds taped but to make sure he asked Arthur to get him some loop bearings.

'I could see the coast line now; a thin light grey streak disappearing into the darkness below and I tried to imagine what the coast would be like. It was a stretch I did not know. I wondered if the tide was in or out. The waves would be breaking on the sand and I also wondered if anyone could hear us. They probably could and might even be talking about us: some people were probably wondering if we were Jerries and might be arguing if it were possible to tell a Jerry by the sound. I began thinking of people I knew and of those at home. They would soon be having supper and were probably listening to the wireless. They did not know I was up here and I was glad they did not know, as they would probably be worrying but it made me feel more alone, knowing they did not know. I felt sorry for those with sons and husbands and friends flying and wondered if it was better for wives to live near the camp with their husbands and know when they would be flying, or to live away from the camp and not know when their men were flying. On the one hand they would have the certain knowledge that their men were out on one particular night-and would be free from anxiety on other nights but, on the other hand, those living away would never know the night their men were flying or were free and would have the continual strain of uncertainty. The coast had disappeared and all that was visible was a misty grey blanket below us with some lighter patches which were clouds.

'Martin took over and did not fly as steadily as the captain. I could see the stars moving slowly backwards and forwards across my perspex as the course altered. The captain told him to watch his course. We were about ten thousand feet up and I turned on my oxygen supply. I told the captain I had done so and he said everyone else had better do the same. I at once began to feel warmer and more comfortable. Oxygen acts on you very quickly: you can be feeling cold and drowsy one minute and almost immediately after breathing more oxygen you feel perfectly normal.

'I once took an oxygen course. Besides being given various medical facts and formulae about oxygen, we were shut into a decompression chamber to test out under the various atmospheric conditions the effects of oxygen and the lack of oxygen. There were six of us in this chamber, which was a large cylindrical structure with chairs and tables inside. Round the walls on the inside were the oxygen plugs and dials exactly as fitted to an aeroplane. From the face mask on your helmet you have a long rubber tube, about four feet long, which you plug in to connect with your oxygen supply. There are two dials near this connection - one marked off in thousands of feet and the other showing the contents of your oxygen bottles. When you want oxygen you turn on a tap which releases the supply and at the same time the flow is registered on the dial which is marked in thousands of feet. You turn the tap until the needle registers the same number of feet as the height at which you are flying.

'Well, we were all sitting round the inside of this iron cage with our oxygen turned on, while the pressure was brought to the equivalent of 25,000 feet. We were each of us given something to do. One man was reading *Punch*; two were writing; two were doing simple sums - in arithmetic; and I had a VGO machine-gun connected electrically with an illuminated target. The target was in the form of a screen, with twelve discs illuminated from behind by electric bulbs. The discs lit up individually and remained illuminated for four seconds, during which time I had to sight them in turn and fire. If I sighted them correctly a bell rang but if not, nothing happened until the next disc showed after four seconds. It was perfectly simple and I was able to do it every time on my trial go.

'The object of this demonstration was to show us how the individual behaves without sufficient oxygen. We were all of us accustomed to the use of oxygen, having used it many times in the air. The effects of insufficient oxygen are similar to those of too much alcohol and people are affected in different ways. Most people become over-confident and some become argumentative; but in all cases their reactions and movements are slowed down and slurred although they may be feeling fine and very confident.

'The man who was reading *Punch* was the first to be experimented upon. At a signal from the doctor - who was watching us through a glass panel from outside - he turned off his oxygen and continued reading out loud. He carried on normally for a few minutes but soon his reading became slower and his speech slurred. He got worse and worse and started repeating himself. The rest of us were in fits of laughter. He was ludicrous; an apparently drunken man with nothing to drink! After a short time the reader became inaudible and started convulsions and the paper slipped from his twitching fingers as he fell sideways off his chair on to the floor. At this point the doctor signalled to us to turn on his oxygen for him. We were by this time more interested than amused and I think each of us was wondering what we should be like when our turn came. I know I was! After our patient started breathing oxygen again we helped him back on to his chair, giving him back his paper and his actions were more or less reversed: he started reading again, at first slowly, but gradually regaining his normal stride. Now, the curious thing was, he had no idea he had even faltered, let alone passed out and he would not believe us until he saw exactly the same thing happen to the next man! Each of us did more or less the same thing, only some were funnier than others: I was told I started off shooting normally but gradually got worse and worse until I was not shooting at all. When my oxygen was turned on I slowly improved until I got back to normal again. I had no idea I had been anything but normal!

'At the end of the course all of us swore we would be very much more careful in the use of oxygen in future and would see that those with whom we flew were the same!

'What is our ETA at the Dutch coast, navigator?' the captain asked.

'Just a minute, sir and I'll find out. Another seventeen minutes, sir' was Bill's reply.

'Thanks. Is everyone feeling all right? How are you in the tail, Riv? Feeling cold?'

'Not too bad, sir. I can't see much, though.'

'No it's pretty nasty, isn't it? It looks as though the Met were right. I hope they're right about it clearing over the target; below was a uniform grey which looked like dirty' cotton wool. We kept flying through cloud which rushed by us as a damp grey haze, making the inside of the aeroplane even darker and it was rather like going through a tunnel in a train. As we passed through these clouds we bumped and tossed like a ship on a rough sea. Above us was even more cloud, with only occasional clear patches through which we could see the stars. The inside of my turret was coated in frost and my oxygen mask was frozen stiff and chaffing the bridge of my nose: I kept on lifting it and tried to put it back on a fresh spot, but each time it slipped back to its old place. I could see about an inch of ice on the top of my gun barrels and I had to keep working my turret every few minutes to prevent it freezing up. It gave me something to do and also helped me in my continual search of the sky. We were not likely to meet any enemy fighters in these conditions, but it would not do to relax.

'On the outward sea crossing one is always more alert and awake. The job is only just beginning and one has the anticipation of what one is going to see or meet but on the homeward journey if all goes well, one is apt to have that feeling of a job done and the worst over and to tend to relax. Also one is tired and it requires more effort to keep alert. A night-fighter pilot once told me that he usually experiences far less opposition from returning enemy bombers than from those who have not yet dropped their bombs.

'We were nearing the Dutch coast, although it was impossible to see it. We hoped to be able to pin-point ourselves when we picked up the Rhine if we were lucky enough to do so. It would be maddening if it was a wasted effort and we would not be able to see anything, after all! Bill said he could see some searchlights through the clouds some distance ahead so we were near the coast, if not actually over it. I felt a sort of tension as I always do when over enemy territory. Anything might happen now! Yes we were definitely over the coast: I could see a glow through the clouds which was obviously searchlights. The clouds did not look as thick as I thought they were. I could see some heavy flak bursts away out to our port, like jewels sparkling: I wondered who was up there and if they were being hit. It was not a heavy barrage and was either being fired at more than one aeroplane or was inaccurate, as the points of light were at different heights and dotted about the sky. Whoever was up there would be giving them a spot of annoyance and trouble below, which was something, anyway! Also a hell of a lot of money was being spent sending up those shells and provided they did not hit anyone, money wasted.

'We were flying at 15,000 feet and the temperature was -30 degrees Centigrade. I was cold but not uncomfortably so. My brain felt slightly numb and it was an effort to concentrate but all the time I felt a sort of inner excitement which prevented me thinking about my physical discomforts. They did not matter, anyway, all I was not there to be comfortable! If I had been comfortable I should probably have fallen asleep. I could feel no sensation of movement except when we were being bumped, which was fairly often. There was the monotonous noise of the engines, which sometimes seemed to change their note. I wondered if the noise really changed, due perhaps to varying atmospheric conditions or if the changed note was only in my imagination. Maybe it was something to do with the pressure on my ear drums.

'Sparks from the exhaust pipes were continually showering behind like sparks from a bonfire in a wind. They and passing clouds, gave me the only impression of speed. Sometimes they came with a burst, as if they had been held back and suddenly released and on several occasions I involuntarily started and realized with a shock that my mind must have been wandering. Surrounded as I was by this grey expanse of cloud and ice; and sitting with perspex and metal within a few inches of me all round - unable to move save only my arms and with the roar of the engines so loud that I could shout without hearing my shouts - my mind felt as enclosed as my body. A few hours ago I was free and was walking and could see people and hear them talk! But now, although there were four men within a few feet of me, their existence felt as remote and unreal as my own body felt; I was conscious of time: one has to-be in the air, as accurate navigation is based upon precision of timing. My thoughts were the same: I was conscious of the-cold and my own discomforts; not that they worried me; they were part of the job. I could think of my home and my friends; of leave; of games; but most of all, of the job in hand. My eyes continually searched the sky - not vaguely, but intelligently and with method - and my hands still worked my turret while all the time my thoughts ran free and seemed to keep pace with our own flight through the air.

'I realized I was still chewing on the same piece of gum that I had started on shortly after leaving base. I unwrapped a fresh piece and exchanged it with the old one: it was frozen so hard that I had to melt it in my mouth for some time before my teeth would make any impression on it.

'The weather seemed to be improving and I could see stars again; also a few searchlights which were playing about on our port: I could see them through the cloud, which was not very thick here. They moved slowly through the air, white shafts of light feeling their way about searching for a victim, like some giant octopus seeking its prey. My spirits rose as the weather cleared; it did not look as though our venture was to be in vain after all.

'Some searchlights suddenly came up from below, where apparently there was no cloud. They seemed to be mostly from behind us, so I called through to the captain but he had already seen them, as the aeroplane leaned over and felt as if it was dropping through the sky. I realized then that the captain must be flying again, although I did not remember noticing the change-over. As the aeroplane leaned over on its turns the searchlights seemed to come at us from our own level instead of from below and to be swinging round. They next looked as though they were falling underneath us and coming up the other side. One of them swung on to us and stayed with us. Gosh! How bright it looked! Every detail of my turret showed up and I felt I wanted to hide! How huge the aeroplane felt and no longer remote and alone. There were men below - and possibly near us in the air too - seeking to destroy us: we were their enemy and they were using their power for our destruction.

'The aeroplane continued its antics; diving and climbing and swinging about the sky. I was very much alert now and strained my eyes against the blinding whiteness of the light. A fighter might or might not be near. Anyway, we were a well illuminated target, should there be one about and I certainly did not intend him to have a sitting shot; and I did not intend him to have a shot at all if I could see him first!

'There was complete silence down the intercom, as we were all too intent to speak. It was as though we had been disturbed by a rude visitor and one that we wished would go! Two more searchlights were on us now and held us despite all the captain's antics. Suddenly I saw a shape above us! It was the silhouette of an aeroplane! Instantly I swung my guns up but almost immediately realized it was our own shadow projected on to the clouds above us by the searchlights below. They went out as suddenly as they had appeared and we were left in complete darkness - a darkness even darker than before and I realized I had been sweating despite the cold. The captain straightened the aeroplane out and continued on his way. I continued staring into the darkness and still chewed my gum.

'Well, you've seen some Jerry searchlights now, second pilot,' said the captain.

'Yes sir.'

'I think that must be our target ahead, navigator. There's a hell of a lot of heavy flak going up.'

'Yes sir. I was thinking the same thing. It's about where it should be.'

'Good! I'm going to climb some more before gliding in. All that stuff looks about our height. '

'OK sir. I think I'll go forward now to the bomb sight.'

This last was from Bill, as he was going to drop the bombs.

'As the captain opened the throttles wider for our climb, a more vicious shower of sparks shot behind.

'I began to feel very excited. The big moment had nearly arrived. We were near our target and soon we would be trying our hardest to identify it. We had had a very clear description from the intelligence officer at briefing and knew just what to look for and I could still see that chalked map in my mind's eye. We were over Germany; over the country with which we were at war! Below were our enemy; the people we were trying to destroy.

'Up there the air was the same as at home. It was cold and clean and unless we were being fired at there was no difference between flying here and flying over England. A few hours ago we were over England; over the country for whose freedom we were fighting and whose people we were trying to save. Although we were over Germany, all we could see was a grey colourless mass with occasionally a few lights and sometimes some very bright lights which were searchlights and some flashes which were guns. Those guns were trying to destroy us and we were trying to destroy the material and men and women that made those guns-and the shells that fired those guns. It was to be a race for one side to succeed. Our skill was pitted against their skill.

'When we get there I'll circle round before going in. See if you can identify the target, navigator. When you do we'll go straight in, we've got plenty of time.'

'Right sir. Another five minutes ought to see us there! Shall I drop one stick?'

'Yes drop one stick if you get a good run-up. If you don't get a good sight, we'll come in again.'

'I was staring out into the darkness. It was a temptation to look at the ground but a temptation that had to be overcome, as the danger of fighters near the target area is always great.

'Yes, that's Cologne all right' the captain said. 'I can see the river. God bless my soul! Look at all those flares! We're not the first by any means. There are several fires already.'

'We were circling round the outskirts of the city and I could now see what was going on. By Jove! What a sight! As the captain had just said, we were by no means the first. I could see three or four large fires glowing red and several specks of red which must have been more fires just starting. There were dozens of searchlights: some appeared stationary, some were waving about the sky, while others intersected at one point. Those I knew held a victim, although they were too far off for me to see him. Those that were weaving about formed intricate patterns against the sky.

'There were bright yellow balls floating gently down: these were flares dropped by other aeroplanes. They were at all heights and by the light of those nearer the ground houses and streets could be seen quite clearly and the river Rhine could be seen, too, as a bright silvery streak winding through the town as the water reflected the light from the flares. Flashing across the ground were bright yellowish-white flashes, which were guns. The flashes were continual and were jumping here, there and everywhere; whenever I looked down they were flashing and leaping about while every few seconds there were brighter and bigger flashes which I knew to be bombs bursting.

'There must have been an unholy din down there, though up in the aeroplane there was only the roar of the engines and our own voices down the intercom. There was plenty to say now and we said it. In the air, shells were bursting with their jewel-like flashes and tracer shells were streaking up like brightly-coloured snakes writhing about the sky. What a sight! Lights everywhere! Beams of light, streaks of light, flashing lights, white lights and coloured lights, searchlights, shells, bombs, fires and flares all at once: hundreds of them all darting about and jumbled together, shooting up and flashing while way up above the stars shone but were forgotten.

'I've got it, sir! No mistaking it.' Bill's excited voice came down the intercom. 'If you turn to the left now you'll see it.'

We turned to the left and the captain closed the throttles. The contrasting silence was a relief as we made our gliding approach over Cologne. I could see brilliant flashes followed by black puffs of smoke as shells burst on either side of us but they were too far off to do us any harm. The turmoil was all around us now, but was not directed at us: we were approaching unheard and unseen and seemed to be spectators high up in some gigantic arena watching this display of fire and fury. Above us; below us in front and behind shells were bursting in a thunderstorm of steel! Searchlights in their dozens were streaking the sky, while tracer shells popped up in their snaky path like little balls of light chasing each other in their climb to the sky. Their upward journey got slower until at last they could go no further and fell down in a gentle curve and then exploded.

'Left, left...steady!.. Left!... steady, steady, steady... Left... steady... Hold it! Steady... Steady Bombs gone!'

Bill almost screamed the last two words. His excitement was infectious and I peered down, straining my eyes against the glow below.

'Can you see the bombs from the tail?'

'No, I can't see the bombs; there are too many flashes but I can see the incendiaries, though.'

They were white and very bright like white Neon lights and were spreading across the ground. Ours were not the only ones: they seemed to be everywhere. Many had already found their mark and had turned into red glowing masses and I hoped ours would do the same. There were many more fires now and they were getting bigger.

'We were still gliding silently and so far had not been touched. We were weaving slightly from side to side, thereby lessening our chances of us being picked up or hit. As we banked on our turns the ground, with all its turmoil of lights and flashes, appeared to tilt from side to side. The captain must have been enjoying himself and was probably muttering 'God bless my soul!'

'When we were clear of all searchlights and flak, the captain opened the throttles: the engines roared their noises again and sparks shot out behind. I felt elated and wanted to sing. What a trip and what a wonderful view! If only all trips were like this one. I had forgotten about being cold or uncomfortable and my heart went out to the captain. By his skill and planning he had brought us safely away from Cologne and danger. I did not mind admitting to myself that I had been dreading Cologne again but I could laugh at all my forebodings now. This would be a trip worth remembering and the sort of trip Leonard would have enjoyed. I felt as though I had really done something though I had done less than anyone, being merely a spectator and had just sat.'

'My joy and high spirits were due partly to the tremendous feeling of relief at the easiness of the trip, as much as to its success. All the forebodings that my imagination had been conjuring up had been in vain: We had left Cologne unscathed, whereas last time we came away wounded and sore. We had seen far more than we had expected to see. There was none of that uncertainty and doubt that one sometimes has on the return journey, wondering where the bombs really fell. Our object had been achieved: we had definitely seen and bombed our target.

I looked at my watch. I would time the flares and note how long I could see them. I had often read in the newspapers and the fires could be seen twenty minutes after leaving the target. Well we had been flying for fifteen minutes and I could see them very clearly. There was a bright red glow on the horizon which tinted the sky above it a wonderful warm red. The glow was still and steady but every now and then was brightened by sudden flashes as bombs continued to burst - probably increasing still more the fierceness of those fires. The searchlights were less bright from this distance and could just be seen as pale silvery streaks which seemed to stay stretched across the sky. The flak bursts were still sparkling and flashing.

'I was reminded of the sunset we had seen on leaving our own aerodrome and thought what a fitting description it would be for the fires I could see and twenty minutes after leaving the target fires could be seen glowing like a sunset. That was just how they looked and I must remember to mention that at interrogation.'

'Have you sent the 'Off target' signal, wireless operator? the captain asked.

'Not yet sir. The wireless has gone for a Burton.'

'D'you think you can fix it?'

'I'm trying to sir. I've got it in bits.'

'Right! Let me know how yell get on. Do you think we are on track, navigator?'

'I don't know, sir, but I hope so.' I can't see a bloody thing!'

'No, neither can I. With any luck we shall be able to pin-point ourselves when we cross the coast.'

'We were passing through some sort of electrical storm. Blue flames were running up and down my gun barrels. The flames did not seem to be coming from the barrels, but to be dancing round and round and up and down, hardly touching them. The trailing aerial also seemed to be alight with a blue flame running its whole length. I was fascinated, watching this flame that did not burn and I called through to the captain to tell him about it. He said the same sort of thing was happening round the airscrews as well.

'It was now forty minutes since we had left the target and there was still a faint red glow away out on the horizon where Cologne was. I could no longer see any searchlights or flak, but occasionally there was a brightening of the glow as bombs continued to burst.

'Arthur called through to say the wireless was still U/S and that there was nothing he could do about it. He had tried everything but could not get a squeak. If Arthur could not do anything, then no one could. The captain said we should have to do without it, then an obvious remark no doubt, but he did not seem to be showing any concern about it. We kept on flying through misty clouds and I could hear the ice cracking on the tail plane and fuselage like pistol shots.

'We should be getting near the coast soon,' Bill said. 'Our ETA is in another seven minutes.'

'There was nothing to be seen; not even searchlights. My excitement had died down and I was once again conscious of discomfort and the cold. I was beginning to feel very sleepy, too: my eyelids felt heavy and I had to strain to keep them open. I began saying the Morse code through to myself to keep awake. What were our beacon letters? 'A:R.' Dit-dar, dit-dar-dit: It would be waiting for us - dit-dar, dit-dar-dit.' They would be waiting for us in the watch office and ops room too. They would probably be getting anxious, not having heard from us as yet.

'Can you see anything from the tail, Riv?"

'Not a blooming thing, sir!'

'Arthur asked if he should drop a flare, but the captain said it was no good, as there was ten tenths cloud below. There was nothing to show us where we were. However, every minute that passed meant we were so many more miles nearer home. I must not forget that bit about the flares being like a sunset.

'We continued on our way, bumping now and then and the minutes ticked by. The engines seemed to take on the sound of our beacon dit-dar, dit-dar-dit. It was as though the engines were calling to the beacon as a horse neighs when it nears its stable.

'We could not say for certain where we were. We might be over the land or we might be over the sea. There was no indication from below: no lights, no signs of land or water; only cloud. There was nothing we could do about it, either. We had come down to four thousand feet and there was still cloud below us. The pilot had got his course from the navigator and was steering it; we knew our ETA for the English coast, which was in about another seven minutes but that was all. Arthur was still working on the wireless set, but said it was pretty hopeless. We could only just sit. Bin could do nothing more either: he had worked out his course and knew his ETA - but without aid from the wireless or without seeing the ground he could do nothing more. The captain was flying with the aid of his instruments and Martin was sitting beside him while I was still in my turret, looking out into the darkness and feeling very cramped and sore in the bottom.

'Our ETA was up, but whether we had crossed the coast we could not tell. We carried on. Suddenly we saw a flare path below us: there was no mistaking it and I think we all more or less saw it at the same moment. There was just a small break in the clouds through which it was visible. We circled and came down through the clouds to below one thousand feet. The captain said he would land there, although we did not know which aerodrome it was, as we could not see the beacon. We rather felt that it was 'a bird-in the hand.'

'It is always rather an anti-climax landing at a strange aerodrome, particularly when one is tired. It usually means a long wait until daylight or until the weather improves before being able to get borne. During the last part of a trip I always look forward to that cup of tea and a pipe before interrogation followed by a jolly good breakfast and then, best off to bed! True, at a strange aerodrome one gets the cup of tea, pipe and breakfast, but not one's own bed.

'The captain was bringing her in now and I prepared myself for the bump. Not that I mistrusted the captain's landing, but I had had too many bruises not to be careful. At the last moment and almost too late, they gave us a red. The captain opened the throttles fully, retracted the undercarriage again and we climbed and circled once more. We saw the sea now for the first time and it was on the wrong side of us! There was no mistaking it. The sea was on the north-west. Surely we could not have crossed England?

'We continued in a wide circuit, with all of us peering at the ground.

'I had forgotten all about fighters. There was the sea again; this time on the south-east! Where the hell were we? We all knew but could hardly believe it: this was one of the Dutch islands!

'We did not waste any time. The captain turned the nose in a northerly direction and kept the throttles open and Bill got really busy. He had got his pin-point! Our feelings were of relief and amazement. Who was the mug who had given us the red? That we shall never know and it still puzzles me. It seems incredible: a British bomber would surely be recognized a few feet above the aerodrome, even at night but instead of enticing us in, they had actually warned us off and never fired a shot. They must have had a mug on duty that night and thank God for it! At any rate, we knew we were over the sea now and Bill said he recognized the island: he gave the captain a fresh course to fly and had some pretty ripe remarks to make about the winds. The captain asked for a new ETA for the English coast; but he said he had no idea, as the wind had completely altered: it had blown us about a hundred miles off our course and was dead against us. We could see the sea quite easily now. Even at a thousand feet and in the darkness we could see the waves breaking into 'white horses' below us.

'The captain said we were getting very low in petrol and unless Bill was wrong about the wind we would probably have to come down on the sea. We had enough petrol for another twenty minutes' flying and about eighty more miles of sea to cross. There was nothing that he could do, except go on flying and fly as economically as he could. It was a question of getting as near to our own coast as we could before coming down. I don't think we quite realized our predicament even then. We had been flying for about nine and a half hours and a lot had happened in that time and we were still far from being out of the wood.

'The fires would still be burning in Cologne, where there would be a lot of suffering and misery. That is what we had intended. Our target had been a large factory and a lot of night-shift workers would have been working there: there would be people dead or dying; there would be people burned there. Some might be alive, living with broken bones, unable to move and with crushed and mangled bodies pressed against them with nothing but the stink of rubble and putrefying flesh for company. There would be people with arms and legs blown off and people with their stomachs blown open and people with half their faces blown away. They might have to wait hours or even days until they were found; unable to help themselves and wishing they could die, yet afraid to die. Some would be badly burnt and would die: others would not die, but would be crippled and scarred always. All these things I had seen when our own aerodrome was bombed.

'While all this was going on we were flying away from the havoc we had caused and would soon be near death ourselves. We were near death now, but how near we could not know. We might, or might not, make a safe landing and even if we did, we might or might not be rescued. We would do all we could and the rest was in the hands of God. Every second we were getting nearer home; and the nearer home we got, the greater were our chances of being saved. We had no wireless, so could not send out an SOS. Had any of our RDF stations picked us up? And were we being plotted yet? Upon this depended largely our chances of being rescued. If we were being plotted we definitely had a chance: if riot, we had practically no chance at all. Still, there was nothing more we could do except wait: we were still flying and we were still alive!

'I had a definite and active job to perform now, as the captain had put me in charge of the dinghy party. My first job was to hack away the fuselage door. The fuselage light was on and I could see what, I was doing quite easily. There was not much room to use the axe and, hampered as I was by the confined space and all my clothing, I soon began to sweat and labour. Yet all the time I felt a sort of satisfaction in chopping with the axe: I felt as a child might feel who had been told he could break his toys instead of being told he was not allowed to! The door came away quite suddenly and was whisked away by the slipstream and I nearly went with it!

'My next job was to see that the dinghy was ready for launching and I wished I had taken more interest in dinghy drill! The dinghy was on the fuselage floor just aft of the door and there was a long cord leading from it and tied to the fuselage. I made certain that this cord was firmly attached as it not only acted as the rip-cord for inflating the dinghy, but was also the only means of securing the dinghy to the aircraft. I called through to the captain to tell him that all was ready at our end and he replied that he could carry on for a bit yet. I leaned out of the doorway as far as I dared and had a look below. The sea was rough; devilish rough: I could see the waves breaking over each other in the darkness. There must have been a hell of a wind blowing.

'Suddenly I saw a light and called through to the captain to tell him but he had seen it, too and was turning round to investigate. It was quite bright and was on the water, but what it was or where it came from I had no idea whether it was on a ship, lightship, buoy, or what Arthur suggested dropping a flare and the captain agreed but we were not high enough for the flare to be of much use, as it touched the sea almost as soon as it ignited and went out. Arthur launched one after the other, but all with the same result: they lit up for a second and showed the sea in all its fury but no sign of the light or from whence it came. As soon as the flare hit the sea, it smouldered and I could imagine the sizzling sound it would make and then darkness. The light was still there and the captain decided to come down as close to it as he could. Bill gave me the Very pistol and some cartridges, which I put inside my flying suit and he, Arthur and Martin did the same with the other cartridges. We now took up our positions ready for the crash-landing.

'Martin and I were lying on the floor right by the open doorway, with our feet braced against what we call the step, which is a raised part of the fuselage about two feet high and just forward of the door. The other two were sitting on the step with their feet by our feet: they had their backs towards the front and we had our feet towards the front and we were all hanging on to ropes slung from the roof. We were coming down now and hanging on for all we were worth. The captain throttled back and we braced ourselves even harder: he must have held off too soon, as he opened up again and then throttled back in a few seconds. There was a terrific crash and the lights went out. We were hurled forward and drenched with icy water and completely blinded by the darkness which was intensified by the sudden change from the light. We struggled to our feet with the water above our knees and the waves crashing against us through the doorway. I groped for the dinghy and hurled it through the opening. I was holding on to the rip-cord and could feel the dinghy inflate and I was surprised at the speed with which it did so. I heard someone shout 'Quickly, sir' and felt the drag of the dinghy against the rope in my hand as I hauled it towards the aircraft. The rope suddenly became slack. It had broken! I hurled myself into the sea and felt the dinghy with my hands. I did not notice the cold. The dinghy was being hurled and tossed about like a cork and I too with it as I was washed across it. [They had ditched off Cromer; 'the lightship' was most likely Cromer lighthouse].

'Arthur and Martin were there too I think. I pulled them on top but Bill was still in the water and I had hold of him by the arm. God! What a weight he was! I was kneeling and hauling on Bill as hard as I could and shouting to the others to help: Arthur half gasped and half shouted back that he could not, as I was kneeling on top of him. So I was! Poor old Arthur. He struggled from underneath me and got a hold on Bill, who was a dead weight and nearly drowned. Martin had hold of his other arm and I had hold of his clothing round his shoulders. He was kicking with his legs and imploring us to pull him with us. We had to get him aboard: that was all I could say or think. We must not lose him; we could not leave him. He was a living body: we were alive and he must live too! The waves were breaking over us furiously and we were being hurled about unmercifully but still we kept our grip on Bill and still we hauled. I was using all my strength and was hopelessly out of breath. Several times we had him nearly with us and each time a wave hit us and we fell sprawling and almost into the sea but still we kept our hold. Arthur, Martin and I on the dinghy and Bill in the water and Bill was drowning and we were getting weaker. Oh God give us strength! The waves had washed us on board so why should they not do the same for Bill? We began to wait for the waves and when Bill was lifted so we pulled and at last - and after how long I have no idea - we had him with us! Bill lay gasping and grinning on the dinghy with us.

'There were four of us huddled together and hurled about but there should have been five. I think Arthur was the first to voice our thoughts but up to now all our energies had been on Bill: we had him in our hands and by our efforts we saved him. But the captain was not with us. We could now see better and we saw him. He was fifty yards away, standing on the fuselage of his sinking aeroplane; the aeroplane in which he had saved us and we could not save him. There he was alone and waiting. We saw him when we rose with the waves and lost him when we went down with them. All the time we were getting farther away and all the time his aeroplane was sinking. What could we do? The answer was nothing, absolutely nothing! We could only watch and thank God for our own lives: we had no paddles and no one could swim in a sea with waves higher than a house. We did not know if the captain had seen us. We shouted but our shout was blown back at and behind us. I don't think he heard our shout and I don't think he had seen us. We watched the captain disappear with his aeroplane and were silent. There were four of us and we were alive. We were· in a sorry plight but we were alive! We were drenched and at the mercy of the fury of the waves and the wind was howling against us with the tops of the waves blown up into a spray and dashed against our faces. Our faces stung, our mouths were filled with water and our eyes smarted from the salt. We were sitting round the edge of the dinghy with our feet in the middle. We realized by now that the dinghy was upside down with all our comforts and provisions underneath and by this time washed away!

'There we were sitting on a circular rubber surface about four feet in diameter which undulated and moved with the weight of our bodies and the rise and fall of the waves. I could feel it writhing underneath me like a live thing.

'We could still see the light with its reflection on the waves and guessed it to be about half a mile away but still we could not tell what it was or where it came from. I got out the Very pistol, which was already loaded and fired a cartridge. The coloured stars shot up and we could see each other clearly by their light: they fell in a curve into the sea and I noticed and was thrilled by the colour and beauty of their reflection in the water. The sea sparkled and shone with all manner of brilliant lights and colour, which suddenly disappeared as the stars were drowned. I fired three cartridges altogether and we looked and waited to see what we might see but we saw no answering signal from a ship or any sign at all. We saw the brightly coloured stars as they soared up and floated gently to the sea and the brightly lit sea heaving and swirling about us and breaking into white clouds of spray but there was no

196

reply or sign from the light.

'We seemed to be part of the waves and were drifting away from the light. Where a boat would have been dashed to pieces, the rubber dinghy merely gave in to the waves and took its shape from them: they could not break it and they could not sink it. The dinghy fitted to our bodies and we stayed with it. We could not hold on, as there was nothing on which to hold but our bodies were relaxed and we leant and swayed with the waves, fitting the rhythm of our movements to the heaving of the water. We were still exhausted from our exertions with Bill and we were sprawled, half-sitting and half-lying, round the edge of the dinghy with our legs tangled in the middle. I looked at my watch, which was withstanding the water and was still going.

'It was five o'clock. We sat and stared at each other and listened to the sea. Martin was sick: he turned his head and vomited with the wind, his whole body heaving. Poor Martin; he felt worse than the rest of us: this was his first trip and what a trip! He was still dazed and bewildered, but he was taking his hardships marvellously.

'Our weight in the centre of the dinghy made a sort of saucer-like depression which was filled with water, so that our legs and feet were immersed. The sea had soaked through our clothing and boots and it was abominably cold. We must bail the water out but how? I took off one of my boots and tried with that, but the leather was so soft and flabby as to be useless. We each had our caps in our pockets and we tried using them: they proved quite good for the job and by holding them open with both hands we scooped the water up and threw it over the side. Only two of us could work at this at once and when the two nearer the direction of the wind tried throwing the water over the side, it merely blew back again. None of us were doing much good, as the waves continually replaced what water we threw out, but at least it gave us something to do and helped us to keep up our circulation.

'I think that first hour was the worst. There we sat in the darkness, shaking with the cold and just able to see each other being hurled up and down and drenched by the waves which at times hit us so hard that they took out breath away; the wind tearing at our clothing, stinging our faces and smarting our eyes, unable to move, or stretch our legs, which were across each other's and already getting numb with the cold. It would have been easy - and it was a temptation, at times - to drop back into the waves which seemed to be stretching up ready to receive us. It would have been over in a few minutes: a few minutes of struggling for breath and choking till blackness and whatever follows. As we were, what would be the end? Exhausted, numbed with cold until we could stay on no longer and were washed away by the waves? Or slow death from starvation and thirst? Or madness? I knew of one crew that had gone insane: they had been in the dinghy for three days and then one by one had gone mad. But we might be picked up! There was always that chance and we held on to it: it gave us strength and endurance. There was that will to live which is stronger than the wish to die. We were young, strong, healthy and had homes and friends. We loved life and we wanted life. Oh dear God, we wanted life more than anything. What had we done in our young lives? Certainly not all that we wished to do. We had been in danger many times before and near death, too: we had seen death but death mothers. Death for us was near now but I don't think I was afraid, I do know that I wanted to live; to see my home and my friends. I loved my home and my friends and I loved life.

'We were very close together in that dinghy. Our bodies were touching but it was more than that: we were sharing our lives and I think our thoughts were the same. If we died, we would die together and if we were saved we would be saved together.

'Our actions continued the same and became almost mechanical, throwing water over the side and keeping our balance on that heaving dinghy. We spoke very little: conversation was in our thoughts. We spoke of our homes and friends but never of our discomforts; they were too real. We spoke of the captain and asked each other time and time again how we could have saved him. How could we?' I still ask myself that question, but have not found an answer yet. I can still see him silhouetted against the sky and not moving, as we rose on the heights of the waves. Could he have reached us? No, no man could have swum in that sea in flying clothing. Bill had lost a boot in his struggling in the sea and we tried to keep his foot out of the water but there was nowhere to put it except on top of our own and the waves wetted it then just the same. It was stiff with the cold and he could not move it. The dinghy was not spinning round, as might have been expected, as there was a canvas sea anchor which kept it straight: Arthur and I were with our backs three-quarters to the wind, with Bill and Martin opposite us. Bill and I did most of the bailing, as from our positions with the wind it was easiest. Martin was still retching and was feeling terribly ill, but he never once complained. His retching and sickness made him very weak and I knew he was feeling wretched. The waves were hitting Arthur and me on our backs and went down our necks and were hitting Bill and Martin in the face. On the whole, I think Arthur and I had the better of it. I still had the same piece of chewing gum in my mouth that I had before reaching Cologne. I suddenly became aware of it and spat it into the sea.

'Our eyes had become quite accustomed to the darkness now and we could see each other quite easily. Martin had slipped more to the middle of the dinghy and was half lying across it with the top half of his back, head and shoulders almost touching the sea behind him: in fact, many times he was submerged but he did not seem to notice it. Poor chap, he was in a very bad way: when I spoke to him he did not seem to hear. Bill was sitting hunched next to him and was pushing his cheerful face to meet the waves. When a particularly fierce one hit him he merely grinned, with the water dripping off his face and all the while he was throwing water over the side. Arthur was leaning forward with his arms across his knees and seemed more cross than anything. Bill and I had to time our scoops for water, as there was not room for both our hats in amongst our legs at the same time: our hands were stiff and painful with the cold and were already swelling and I found it difficult to hold my hat.

'At times I felt in the depths of depression and despondency and as if a heavy weight were on my chest. At these moments I needed every ounce of courage I possessed to keep my body composed. The others were probably feeling the same - I know Martin was - and I hoped I was concealing my feelings as successfully as they were. At other times my thoughts were almost gay in comparison and I felt a sort of thrill and exhilaration in our wild gyrations: my spirits seemed to rise and fall with the waves as they took us with them where they would. We were never still: up and down; at times so high it seemed we should be blown from the tops. At other times so low it seemed the weight of the water must come down, swamping and drowning us. But our dinghy, the most seaworthy craft of its size, was always on top.

So we sat and watched the waves get lighter, lightened by the dawn which came with storm and cloud and rain while at base they would have given us up by now and chalked up one word - Missing - against our names. I had seen that word many times and had often wondered where those men would be. They would be wondering that about us now, back at base: we were just one of those crews that Failed to return. Martin spoke for the first time and said: 'When will they send telegrams to our next of kin?' 'Oh, when the old adj starts work; not for another hour or two yet, anyway they always wait a bit to give us a chance to turn up,' I answered. 'You're married, aren't you, Martin?'

'Yes sir. I'm married.' 'Where does your wife live?'

'In the village; I live out.'

'I expect the adj or padre will go round and tell her,' I replied.

'Much better than having a telegram. I always think a telegram is much too brutal and seems final, somehow. You'll have to cheer her up to-night, Martin.'

'So you really think there is a hope, sir?'

'Good lord, yes, of course I do! They'll be scouring the water for us by now.'

'I've got a date in York tonight,' said Bill.

'I wish my girl lived in York,' Arthur replied.

'And all the time it was getting lighter. What should we see?

'The sky got lighter and the clouds took shape: dark clouds racing low across the sky and competing in fury with the waves; waves as fierce as ever, but majestic in their beauty. We could see their colours now; colours changing all the time.

'You can't describe colours; there are too many - any more than you can describe music. Music does not consist of single notes any more than colour consists of one colour: music is a harmony and rhythm of notes, while colour is a harmony and continual change of an infinite variety of subtleties of colour. You can say a thing is blue or green or red or brown: but there are hundreds of varieties of blues, greens, reds and browns and some blues are so nearly green and some browns are so nearly red that the changes from blue to green, or red to brown, are so gradual that it is hard to say when blue ceases to be blue and becomes green, or when brown ceases to be brown and becomes red. So it was with the sea; never still, never the same - always changing in shape and colour.

'There was no sign of ship or land. We were alone and waiting and we wondered just where we were. Bill said he thought we were thirty or forty miles from our own coast and we began wondering if we were on any shipping route. It was too early yet to expect help, as it was only just light and they would not send out anything until now.

'I wish I had a cigarette!' Arthur said. He brought out a packet soaked to pulp and threw it into the sea. It was carried away and disappeared instantly.

'We saw a speck below the clouds: it was an aeroplane flying towards us! We watched it excitedly and intently as it got nearer. I got the pistol out ready to fire: it felt heavy and awkward in my numb fingers. The aeroplane came nearer, flying fast and low and we saw it had twin engines and a single rudder, so we took it to be a Blenheim.

'I waited until it came nearer and fired the pistol: I found the trigger hard to pull and had to use both hands, my fingers were so weak. As I fired, Bill shouted 'My God! It's an 88!' He was right; it was. Thank God it had not seen us, but carried on its way and disappeared into the clouds.

'He's been up to no good,' Arthur remarked. Very soon some seagulls appeared, apparently from nowhere. They hovered around us, motionless in the wind, with only their heads turning from side to side. 'Let's hope they stay and bring some pals,' I said. 'A ship might see them if it can't see us and guess there's something near!'

'It is amazing the way seagulls will appeal; when there is a ship or human being on the sea. They obviously hope for food. These gulls soon found they had come to the wrong place and that we had nothing for them and they continued on their mysterious journey - we knew not where. They had made us feel less atone and had somehow given us hope.

'I was struggling to reload the pistol, but the cartridges were soaked and swollen. Eventually, by peeling off some of the outer casing, I managed to push one home and close the breech. It started to rain hard and we leant back our heads and let the water splash into our faces and trickle down our throats. At first it was refreshing but soon it became monotonous and uncomfortable as it poured down our necks, making us even wetter than before. We sat with shoulders hunched and Arthur swore.

'The rain did not last long though and the weather showed signs of improving. The clouds were lifting and there were some breaks in them: a patch of blue sky showed overhead, giving more colour to the sea. More seagulls kept appearing; some staying and others flying past without taking any notice of us. I saw some gannets fly by, with their pointed wings and heads, on their straight course, just missing the tops of the waves: they seemed to be flying with more purpose than the other gulls. I love watching gannets and, even there I could watch them with interest and pleasure. There were oyster-catchers too, chattering in their high-pitched voices as they passed with rapid wing-beats. I was surprised to see them so far out to sea, as I had, always imagined them to be shore birds. Perhaps this meant that we were nearer land than we thought? I confided my hopes to the others.

'Again we saw an aeroplane approaching in the distance! There was no mistaking it this time, with that short fuselage and twin rudders. It was a Hudson, flying about 200 feet above the sea and about half a mile from us! I struggled frantically to fire the pistol but my fingers were so cold and stiff and swollen as to be almost useless. Bill was helping and at last we got it off but too late, as the Hudson was well past us and went out of sight. I had even more difficulty in reloading the pistol and while I was still trying the Hudson reappeared, this time on the other side of us. I tried my hardest to load the pistol while the others waved frantically but in vain! The Hudson disappeared.

'I remarked that if there was one there would probably be others. We felt better. The clouds were lifting and the sun threw its rays horizontally across the sea and dazzled us: the lights sparkled and danced on the broken surface of the sea that shone as brightly as the sun. The wind was still tearing at us and the waves were still soaking us and we were shaking with the cold but we felt more cheerful with the light. I had never been so cold before. The others' faces were white and patched with blue and I could see their teeth chattering: my own sounded like a roll of drums! My legs were stiff and painful and my back ached: I would have given anything to have been able to lie down. I tried to change my position and move my legs, but found I could not. I picked up one and moved it a few inches. The joints hurt abominably when I moved them, but I could feel no sensation in my leg at all. The only one oblivious of our surroundings was Martin, who lay back and stared upwards at the sky and I am sure without seeing the sky.

'I wonder if she knows yet' he said.

'Will she take it very hardly, d'you think?' I asked. I knew he was referring to his wife.

'I don't know what she'll do. She's all alone.'

'The adj will look after her,' I said. 'He's very decent like that. Anyway, you'll be with her yourself soon.'

'I told her I should be back before dawn.'

'Well you'll be a bit late but think what a welcome you'll get. It'll probably be worth it.'

I did not like the look of Arthur, He had not spoken for some time and his eyes were wandering round without looking at anything. He asked me the time.

'Getting on for nine o'clock,' I told him.

'I can't stand another four hours of this! Have you got your revolver, sir? '

'No - I've lost it,' I answered him. I was carrying it tucked in my flying boot and it came out in my struggle on to the dinghy.

'Arthur was trying to open the clasp-knife which was slung round his neck but, thank God, his fingers were too weak from the cold for him to be able to do so.

'Don't be a damned fool, Arthur!' I said. 'Go over the side if you like but you're not going to make a mess in here!'

He began to laugh and soon said: 'I'm sorry sir.'

'Nothing to be sorry about,' I answered. 'I knew you were only fooling.'

'I wasn't fooling.'

'Oh yes you were! Besides, what about your girl? She would have minded, you know.'

'I told Mum and Dad I'd finished with operations,' Arthur said a few minutes later. 'I wonder what they'll do when they get the telegram?'

'Probably give you a good raspberry when you get back,' I answered.

'Our eyes were continually searching the sea: when we were carried to the top of a wave, Bill and I would stop our bailing and strain our eyes to the horizon. Our range of vision was very limited, though, as rolling, monstrous waves prevented us from seeing far. We did once see a ship when we were lifted by a particularly high wave but only for a second or two and then it was gone and we never saw it again. We tried to see if we were drifting in any particular direction, but there was so much movement all around us that it was impossible to tell.

'I wonder why they gave us that red?' Bill remarked.

'I can't imagine but I'm jolly glad they did!' I replied.

'My gosh, so am I!' Bill said quickly.

'I wonder if I'll ever see my DFM?' Arthur said after another silence.

'I didn't know you'd been put up for one,' I exclaimed. 'Jolly good show!'

'I'm not supposed to know but Jimmy told me.' Jimmy was his last captain. 'Do you think we shall die, sir?' he went on.

'No, I don't think so' I answered. 'No I don't think we shall die. It'll be grand to look back on though, won't it?'

'I'd rather not look back on it!'

The clouds were nearly all gone and the sun was getting higher in the sky. The sea was not so rough and we were not getting swamped quite so often. Our bailing was becoming of more use too but each time we got nearly to the bottom; a fresh wave came in and gave us more work to do.

'If this was peace-time and we were near the shore, we'd be paying money for this sort of thing,' I remarked, after a long silence.

'I don't mind if I never see the sea again,' Arthur replied. 'My home's by the sea,' I told him.

'Where?'

'Cornwall. D'you know Cornwall, Martin?'

'No, sir, I've never been there.'

'You ought to take your wife there when you get leave: they're sure to give us some leave after this.'

I was getting terribly weak and stiff in the joints and it was becoming a real effort to throw the water over the side: both Bill and I kept stopping in our movements and we were getting slower and lower. My legs were completely numb and it was impossible to move them except by picking them up: on several occasions I picked up Arthur's foot, thinking it was mine! Bill's foot without the boot would be in a bad way by now. He said he had no feeling in it at all.

Arthur suddenly produced a small packet of chocolate, which he divided into four. We ate it greedily, in spite of the fact that we should probably feel more thirsty. Martin was sick again.

I looked at my watch. It was nearly eleven o'clock. How much longer could we last? We could last until dark but I doubted if we would be able to see through another night. The last two hours' seemed to have gone by very quickly.

'They must have sent the telegram by now!' Bill remarked.

'Yes, I expect they have' I answered. 'It seems strange to think of what is happening at home. We know we are safe, but they don't. It's worse for those at home: if we die it's the end but for them it's only the beginning. But don't let's talk of dying! It's only eleven o'clock and there's bags of time for them to find us yet.'

'We began discussing and imagining how they would set about it. Would any of our own squadron be searching for us? Probably they would. The sky was quite clear of clouds now and the visibility was good. If only the wind would drop! It did not give one a chance to get warm: it seemed to get right inside our clothing and into our bodies.

'Arthur had dropped forward with his head between his knees. I did not know if he was asleep or in some sort of coma but I was glad. I hoped he would lose consciousness and stay unconscious: if he had to go, I hoped he would not know the end. Can you tell the moment of dying? If you are shot through the heart and die instantly, can you tell the instant of dying?

'Is dying like sleep? Oblivion? Do those who die in their sleep know any difference between their sleep and their moment of dying? Is there blackness for ever more or does your mind re-awake in a different body and different surroundings? I began wondering these things. Should we know when we were dead? If we lost consciousness, would that be the end? Or should we know in our unconsciousness that we were dying?

'Bill and I looked at each other and each seemed to understand the other's thoughts: we seemed to gain strength from each other, too. We were still bailing water out, but very slowly and we would continue to do so until our arms could no longer move.

'Suddenly Bill said: 'Listen!'

'What is it?' I asked.

'Can't you hear? An aeroplane!'

'So it was and a Blenheim this time, all right flying about two hundred feet and zigzagging about. Oh, God make it see us! Arthur was alert now and we all eagerly followed its flight with our eyes, turning our heads first one way and then another as the aeroplane changed its course. Its snaky course was getting nearer; very slowly, but nearer as we watched it with alternate waves of hope and despair: turning towards us and then away again. It was very close now and surely must see us. I had the pistol and was doing all I could to fire it but I could not! The others tried, too, but with no better luck: our fingers just would not move. And all the time the Blenheim searched; searched for us, but did not see us. It was moving away now and had not seen us: it was still searching but it was going away. We watched it turning first one way and then another but always away from us: we watched it going away and we watched it until we could see it no longer. It had gone and we were alone and we felt more alone than before!

'We continued to watch the sky when the Blenheim had gone and we saw it reappear, still flying from side to side. It seemed to be making straight for us this time and would surely pass overhead. It was getting very near now and we waved our arms frantically. We shouted in our frenzy as though the anguish of our shouts and minds must reach those men searching a few hundred feet away. Ever nearer the Blenheim came, turning this way and that. Suddenly it turned and flew right around us, less than a hundred yards away. It had seen us and the crew were waving to us.

'Can you imagine our joy? I don't think so; unless you have been facing death for nearly eight hours. We felt as a condemned man must feel who is reprieved on his way to the scaffold. I felt a surge of happiness such as I had never known before. Although our bodies were numb and stiff, our minds recovered instantly. I looked at my watch. It was eleven-thirty and we had been in the dinghy a little over seven hours.

'Our hearts were filled with joy, which showed in our faces. We had been found! We were not actually safe yet but we had been found. We were no longer drifting aimlessly about the sea and getting weaker but had been seen and were being cared for. We had been spotted and were now the centre of focus of a vast organization working for our safety and homecoming. The Blenheim would have wirelessed to its base, giving our exact position; the Navy and air-sea rescue service would be informed and ships would be sent speeding to our rescue! We could look to the future now with joy, instead of with sadness and dismay. We began speculating as to what would pick us up. Would it be a destroyer or a motor-boat or would they send a seaplane? I knew a crew who had once been picked up by a destroyer. Could you get a hot bath on a destroyer? They would be sure to have hot meals, hot tea, hot whisky and a bed! Yes, a bed perhaps, best of all, a bed; a warm, comfortable bed!

'We could now think with pleasure of such comforts whereas before we had not dared to speak or even think of them. How long would it take for help to arrive? I guessed about three hours. We were already craning our necks and straining our eyes in the direction of the shore looking for the ship to take us home, in the same way as people on a railway station rook down the line for a train that is late.

'Round and round the Blenheim flew, watching us; guarding us. We made a mental note of its number. We must write to the crew when we got ashore; go and see them and thank them. Had they any idea how we felt? Could they possibly know of our gratitude and joy? How could they? To them we would be no more than just an unfortunate air-crew that had not quite made it: they would be glad that we were safe, but they could not be more than that. We were just four airmen whom they had saved and they would be glad but that was all. But to us they were as brothers: they had saved us when all seemed hopeless and lost.

'We owed our lives to them; now could we let them know of our appreciation and happiness? They, too, must share our tremendous joy: they had made us happy and they should know of our happiness. We waved to them and thanked them in our hearts:

'Bill and I continued our bailing for some purpose now. Whereas before, our actions had been purely mechanical, now they were essential! We noticed for the first time that the dinghy was not quite so buoyant and was slowly losing air. This caused us some concern: how long had it been deflating? The buffeting it had been having was beginning to tell on it and there must be some tiny punctures or cracks. We tried to see where the air was escaping but could see no signs of any holes. We sat as still as we could, hardly daring to move for fear we should make the leakage worse! Some horrible thoughts began to haunt me. Suppose the dinghy was more damaged than we thought and was ready to split! It would be appalling if it sank now and we were drowned before the ship arrived. Until the Blenheim came it would not have mattered but now it did not bear thinking about. The dinghy had had some terrible blows from the waves, which must have strained the rubber to its utmost endurance. There was still a big swell on and although the waves were only breaking over us occasionally now, the dinghy was being twisted and contorted continuously: there were great folds across the rubber and I could feel and hear it chafing with every movement of the sea. How much longer could our little craft stand up to the strain and how long had the Blenheim been with us?

'Forty minutes. Pray God we could hold out until a ship arrived.

'Bill suddenly exclaimed: 'I can see a ship!'

'Where? Are you sure?'

'Yes. I saw it a few minutes ago, when we were on the top of a wave.'

'I turned my head in the direction towards which Bill was pointing and sure enough, when we rose high, I could see a speck in the distance which was obviously a ship. If it was really coming for us, it was coming far more quickly than we had ever dared to hope. Bill kept watching it when we rose with the swell and he reported that it was getting nearer and that he thought it was a destroyer. Presently he said he could see two more. Magnificent! Three ships for four men! We began discussing whether we should get leave and how much. It was marvellous to be able to look to the future and make plans again: I could contemplate home and friends now without a pang. 'What a lucky chap I am and how good life is' I thought.

'It was now possible to discern details on the ships as they came pitching and tossing towards us. We could not yet see what type of ships they were: they might be destroyers, or they might be some smaller craft but what did it matter? They were ships and they were coming for us; coming to take us home. Very soon we would be warm and cared for. The Blenheim was still circling us and evidently was going to stay with us until we were aboard the rescue ships. Every second brought them nearer and all the while our joy was increasing. No longer were we four destitute airmen but four lucky men whom God had decided to save and who had lives to live.

'The ships were not the destroyers we originally thought they might be but they looked like some sort of trawler. We could just make out figures on the nearest one. They were closing towards us very rapidly and our joy and excitement knew no bounds. We started waving to them, but could not see yet if they were waving back. As the leading ship came within hailing distance, one of the men in the bows called out: 'Is anyone hurt?' Then almost immediately on our reply that we were all right: 'Have you been seasick?'

'As they drifted slowly nearer, questions and replies were shouted across the water. If our joy had been great before now it was terrific. If only the captain had been with us, life would have been perfect.

'Men were leaning over the side watching us, ready to help. One of them threw a rope across, which we managed to hold between us and we were hauled alongside the trawler with our dinghy rubbing against its friendly bull. We were rising and falling with the swell; sometimes almost level with the deck and at other times right down by the keel. One of the sailors lowered himself down the rope ladder into the dinghy to help us in being hauled aboard: our legs were useless and would not hold the weight of our bodies, so we could not climb the ladder but by much pushing from below and pulling from above we were at last dragged aboard the ship and lay helpless on the deck.

'We could only smile and thank our rescuers time and again. They put lighted cigarettes between our lips, gave us some neat whisky and pulled off our flying clothing. They then helped us below into their cabin, where there was a roaring fire and undressed us: we sat naked in front of the stove absorbing the glorious heat, while the sailors raked out blankets, stockings, woollen pants and jerseys.

'Very soon we were sitting round the table wrapped in blankets, with mugs of hot tea, glasses of whisky and plates of bacon and eggs before us talking and laughing and joking. Never before had life been so good or people so kind. Never shall I forget those men, rough fishermen used to daily hardships and dangers themselves; yet as kind and gentle as mothers. All our wants and comforts were attended to with the greatest care and thoughtfulness: our mugs of tea and glasses of whisky were never allowed to remain empty; we were given cigarettes and also some tobacco for my pipe; bunks were prepared for us and we lay talking and smoking until we fell asleep, swaying with the regular roll of the ship. Life was too good and full of interest for us to sleep for long, though. By sleeping we missed the full thrill and enjoyment of being alive! I kept reminding myself that I was alive; that there was still the future.

'We told our stories time and again to fresh sailors coming to see us. We kept them plied with questions, too: how had they found us? Where were they taking us? Where had they come from? And so on as the hours passed by: hours of happiness and relief. When we neared port and the time was approaching for saying good-bye to our new-found friends, we exchanged addresses and wanted to give the skipper five pounds to split amongst his crew, as we could just scrape up that amount between us but no, nothing would induce any of them to take a penny! All they asked for was a souvenir, such as a button off our tunics. I told the skipper to get busy with the scissors, which he did!

'We slept that night in hospital, where incidentally I think Ming got more attention than I did and before going to bed we telephoned our unit and homes to let them know that we were safe.'

There was no such call from Squadron Leader Florigny's younger brother, Pilot Officer Alan Addison Florigny. By a cruel twist of fate he and his crew had perished, one of the five Whitleys that failed to return. A Wellington was also lost and fourteen further aircraft were abandoned and crashed in fog over England with a dozen crewmembers killed; two missing and five injured. The two brothers are both commemorated on the Runnymede Memorial.

'To cut a long story short - and a still longer journey by rail and road' recalled Rivaz 'we arrived back at Topcliffe aerodrome at about midnight on the following night. We had hoped to be allowed to go straight to bed, but on arrival we were greeted with a message to report to the Station Commander at the ops room. I glanced behind me just before going in and saw Bill with a broad grin on his face. 'Don't look so damn fit, Bill, or you won't get any leave! Start trembling or something' 'I whispered. Two days later we went on leave.

'When I returned to the squadron from leave, I went to see the CO [Wing Commander Francis Charles Cole]. He told me that Bill, Arthur and Martin had been sent on a rest and he asked me what I would like to do. I wanted to know if my nerve had been affected and the only way I could tell was to do another trip. Bill and Arthur had both done considerably more trips than I and both needed and deserved a rest but I felt I would like to go on longer yet. I asked if I could do another trip as soon as possible and would the CO take me? He said not that night, as he was already fixed with a gunner, but next time.

'That night [7/8 March] he and his crew were missing.[62] I put in another request to be posted to the Halifax squadron that Leonard Cheshire was on [35 Squadron at Linton-on-Ouse] and much to my surprise and joy, my request was granted immediately and I moved the following day. I arrived at a slack time for the squadron and spent the period learning all I could about the new aeroplane. I had hoped to join Leonard's crew straight away but for some obscure reason this was not to be and I was put to fly with my flight commander, Braddles. He was a good pilot, but I was very disappointed not to be allowed to fly with Leonard again. Both Leonard and I put in repeated requests that we should be together but in the middle of our negotiations he was sent to America for three months. So that was that!

'Braddles used to vary his crew somewhat, with the exception of Wheeler - his engineer - and myself, his tail gunner. Wheeler was a quiet lad of about twenty and a first class engineer. He was very quiet and precise in his movements. All his actions were deliberate and he gave the impression that he would do nothing without thinking about it very carefully first. Even his smile, which started with his eyes, seemed to be carefully contemplated before it was allowed to appear.

'The first trip I did with Braddles was to the Ruhr: then Magdeburg, Hanover, the Ruhr again and so on. On several trips we carried the new four-thousand-pound bomb; a monstrous, ugly thing which looked more like an engine-boiler than anything else. When it was released, the aeroplane seemed to give a sigh of relief and rise as light-heartedly as a lift. On one occasion I missed this light-hearted feeling in the aeroplane as the navigator said-'Bombs gone!' - and sure enough, a few moments later I heard him say - 'The bloody thing's still there!' He tried everything he could to free it, but without success, so we had no choice but to bring it back with us. Normally it would have been perfectly safe to land with it but we did not know how near it was to dropping off. Also we had been hit by flak over the target and we wondered if our wheels or undercarriage might be damaged. It was an unpleasant moment as Braddles made his approach for landing but I consoled myself with the thought that if it did go off we should not know!'

When Richard Rivaz returned from leave and got back to 35 Squadron, the first person he saw was Leonard Cheshire, dressed in his roll-top jersey, just getting ready to set off for Berlin.

'Congratulations, Revs!' he said.

'What for, Leonard?'

'For shooting down a fighter.'

'Well, it was either him or us. Who've you got in the tail?'

'Martin.'

'Oh, he's all right, but I wish I was coming with you.'

'Next time, Revs.'

'The next time was not Berlin; it was Düsseldorf but the one after that was Berlin and I was with him. I was glad he was taking Martin, chiefly for Martin's sake. He had not flown for some time. The last trip he was on he was wounded by a cannon shell, which exploded in his turret. They were being attacked by a fighter and the shell that wounded Martin badly damaged his turret, making it very difficult to operate. However, he managed to work it somehow and shot the fighter down. While he was in hospital his pilot was posted and I knew that Martin was not happy about flying with just anybody: I felt that Leonard would give him the fresh confidence which he badly needed.

'That trip I did to Berlin with Leonard was one of the best I have ever done. Everything was perfect. The moon was full and we were able to map-read our way across Germany. I had never seen the ground so distinctly at night before. Fields, rivers, lakes, railways, roads; all showed almost like day. I was even able to distinguish cornfields. The suburbs around Berlin appeared, getting gradually thicker until we were over the city itself. We cruised around almost like a 'Cook's tour' and I could see the houses, streets and parks as clearly as though it was looking at an aerial photograph. The moon was so bright that the searchlights were almost ineffective in their attempts to compete with its brilliance. The guns left us alone that night, too. Whether the shells were not bursting at our height, or whether we were just lucky, I don't know: anyway, we got off scot free. After we had dropped our bombs we cruised around a bit more before turning north on our long journey home!

'About this time we had a new CO - Wing Commander Robinson - who was known to most people as 'Robby'. He was happiest when he was in the air and he did not waste time getting there, either! I flew with him on his first trip with the squadron.' Basil Vernon Robinson was a pre-war pilot, born in Gateshead in 1912 and initially commissioned in the RAF in 1933. He quickly distinguished himself playing Rugby as a wing three-quarter for the RAF and his home county XVs and by April 1938 had been promoted to Flight Lieutenant. By 1941 he had risen to Squadron Leader and been awarded a DFC in July of that year during an operational tour on 78 Squadron. An ebullient character, with a distinctly unorthodox approach to certain RAF customs and procedures, his trademark was a generously proportioned ginger bushy moustache. Robinson had joined 35 Squadron in late 1941, at Linton-on-Ouse.

As Richard Rivaz entered the briefing room, which was rapidly filling up with air crews, he looked at the map eagerly with the others to see where their target was. 'Our route was marked by a piece of red-tape which terminated at Cologne. Cologne! I suddenly felt empty inside and weak. What would happen this time? But perhaps the third time would be lucky.

'It was a pitch dark night as we circled dropping flares, searching for our target. The flares showed fields but that was about all! We were one of the first due over the target, so there would be no fires to guide us. There were fires though but most of them were very obviously dummies. They were too even and without smoke or else they were not bright enough or perhaps too bright. A genuine fire looks very red from the air and sends a glow all around. It is not a consistent mass of red either, but it flickers and changes colour and there is always smoke.

'We continued searching for some time. Both Robby and the navigator were convinced we were near Cologne but there were no searchlights, no flak, no-one else's flares. We must be early or else the navigation must be wrong. Suddenly Cologne awoke and showed herself some distance away on our port. Her hundreds of guns and searchlights all got busy together. There must have been several aeroplanes over there too, as flares like yellow balls were mingled with the flak and the searchlights. They seemed very unconcerned and aloof from the hubbub around them as they floated proudly and majestically down.

'My God!' Some poor bastard is getting it' said Robby. I was watching. About a hundred searchlights were all pointing at one spot and this point where the beams intersected was alive with sparkles and flashes. It was iridescent and writhing with brilliance as a piece of bad meat moves with maggots. Robby was right: some poor bastard was getting it and getting it badly. Every gun and searchlight in Cologne seemed to be concentrated at that spot at that aeroplane waiting to drop its bombs.

'A red glow appeared amongst the flashes. It divided into two glowing patches which were falling, still held relentlessly by the searchlights. One of the glowing masses exploded and disintegrated into many more glowing pieces! Still the searchlights followed the burning wreckage followed it until the beams were flat with the ground while flares hovered around and bombs flashed below.

'Whew!' I heard Robby exclaim.

'We were over Cologne, searching for our target! Searchlights were making it difficult to see: there were so many of them and their beams threw a protecting screen of lights across the city. They were weaving about the sky looking for another victim: they seemed uncertain as they groped cautiously about the darkness with their long tentacles. Who should it be? Who would they choose? They seemed to be gloating over their last victim and prolonging the impending moment of their next feast. They decided and were moving in our direction! Some hurried as though eager to be there first. Others lingered as though not quite certain; or else prolonging the delicious moment. They came on us singly in pairs and then in dozens. With them came the flak, bursting everywhere and without pause. Shells above us, shells below us, shells behind us and on either side. Their noise was deafening and drowned the roar from the engines. They stayed with us without effort as we turned and twisted, trying to free ourselves from their villainous grasp. The searchlights dazzled and mocked us as they clawed and pawed with their evil clutches about our aeroplane, trying to snatch them from the sky. The guns were getting angry with them and with each other and with us in their vain attempt at our destruction.

'The searchlights were smug and conceited as they fondled us in their filthy fingers. They were holding us steady for the shells to hit but the shells were not hitting hard enough! They must come closer. Look, the searchlights were touching us; holding us, holding us tightly in their death grasp! They must come closer, must tear right through as they did to their last victim. Think of him how he had gone down burning, with the bodies inside struggling to free themselves! I was completely blinded in my turret. Everywhere there was light; shafts of white and violet light, relentlessly rude and enquiring. The light was around my turret, taunting me, infuriating me. I felt I was exposed to hundreds of inquiring cool and piercing eyes!

'A shell splinter crashed, tearing through my turret, ripping a great hole in the side. I seemed to be filling all the turret and I wondered how it missed me. I wished I was smaller and could hide. I felt so exposed and visible, sitting stuck out at the back of the aeroplane surrounded by perspex.

'Robby was singing and letting out periodical whoops and war cries. Suddenly he said 'I've got it! I'm going right over it. Get ready navigator!'

'About a minute later he called through to me: 'Could you see them burst, tail gunner?'

'I had not seen a thing, only light. We had done our job and were free to leave. The searchlights left us as they had come on to us, singly and in numbers. We were clear at last and flew once more through the darkness, which seemed even darker than before. The searchlights were still busy, but not with us. They had found another victim and were tormenting him.

'As I watched them getting fainter, a fresh batch shot up from below. There was no indecision about this batch: they were on us straight away with a determined, vice-like grip. They were unaccompanied by flak with an obvious reason ... fighters!

'There were as many searchlights as before and as bright. All I could see was brightness and reflections on the perspex. Was there a fighter behind us? Was he stalking us, getting nearer and nearer, waiting his chance to let fly? I strained in to the dazzling, bright light, turning my turret this way and that.

'Suddenly it came! I heard his guns first and saw red tracer bullets coming at me. As I heard his guns I opened fire, shooting at the flashes that were all I could see of him. I don't know what happened. I only know that as I fired, he stopped firing. It all happened in about three seconds.

'Are you all right, tail gunner?'

'Robby had seen the tracer shooting past us and he imagined I had been hit, as I was silent. In my excitement I had forgotten to switch on my intercom.

'As we crossed our coast it was getting light and I was discussing with the navigator the prospects of the wild fowling on the marshes below.

'Stop talking all that cock! We've had enough shooting for to-day!' Robby interrupted.' [63]

Endnotes Chapter 9

62   Wing Commander F. C. Cole and his crew and Whitley V Z6468 were lost without trace on the raid on Kiel.

63   On the night of 18/19 November 1942 the target was Turin involving a long haul across France and the Alps. Piloting Halifax 'S-Sugar', Robinson's outward trip was uneventful and he duly reached Turin and bombed, then turned for home. Unbeknown to the crew one of the marker flares had hung-up and was still in the bomb bay and as the Alps loomed into view the Halifax fuselage was suddenly filled with acrid smoke as the flare ignited in the belly. Robinson quickly realised the cause and ordered the bomb bay doors to be opened, hoping the flare was loose enough to drop away, but the exposed flames merely spread to the port wing and increased in intensity. Robinson ordered his crew to bail out immediately and each man left the apparently doomed Halifax in quick succession and then the fire went out! Robinson was alone, in a damaged Halifax and about 700 miles from base; a prospect of four solid hours of piloting an aircraft without benefit of his crew. He thought of bailing out but he decided against it and he crossed the Alps and then France. On reaching England he had no means of calculating an accurate route to his own airfield. Without a radio to warn those below he put the Halifax down on the first aerodrome he could find, at Colerne in Wiltshire, switching on all lights he could reach in order to indicate his presence and intention. Rolling to a stop, he climbed out to meet the ground staff awaiting him and derived embarrassed amusement at the looks on their faces when they realised that no other crew men were aboard. His lone flight brought Robinson a Bar to his DFC. On 1 May 1943 Robinson was promoted to Group Captain and he was appointed commander of RAF Graveley. Station commanders were severely restricted by higher authority from undertaking operations, being rationed to two or three sorties per month at most. Robinson, typically, selected particularly tough targets for his few trips and on the night of 23/24 August 1943 the target was Berlin. He personally briefed Graveley's crews and then took a scratch crew out on 35 Squadron in a Halifax. Of the 719 bombers dispatched to Berlin that night - 117 of these PFF aircraft - 56 failed to return, including two 'Station Masters': Group Captain A. H. Willetts at Oakington and Group Captain Basil Vernon Robinson DSO DFC* AFC.

Flight Lieutenant R. C. Rivaz DFC was killed on 13 October 1945 when the Liberator VI transport he was flying in crashed on takeoff at Melsbroek in Belgium. The Liberator which was piloted by Flight Lieutenant Peter Green, was carrying a crew of five and 26 passengers; members of the RAMC including 223 Field Ambulance, a dentist and a member of the catering corps. It was found that the aircraft had been incorrectly loaded. The 37-year old Rivaz was buried in Brussels Town Cemetery. At the time of his death he was researching another book, on RAF Transport Command.

Chapter 10

Hampden Ops

I would express my complete confidence that with six Hampden squadrons, including one long-range squadron for the Eastern Baltic... and an adequate supply of these weapons, given reasonable weather conditions we could close all the ports that Germany is using for her overseas ventures and keep them closed in a manner which will either defeat these ventures or cause her most tremendous losses in an effort to continue.
Arthur Harris A.O.C 5 Group, in a letter to Air Marshal Sir Edgar Ludlow-Hewitt, Inspector-General of RAF, 17 January 1940.

On a dark night in December 1940 Gilbert Haworth crouched over his navigator's table in the nose of the Hampden bomber as the crew sped through cloud on the way home from enemy territory. Haworth had joined Bomber Command as an air observer in April 1938. His Hampden could not rise above the cloud, because, as he recalled 'it was unbroken up to 18,000 feet and we could not get below because the base was down to 2,000 feet. Our decision was to stay at 15,000 feet until about 20 miles from England and then we would descend steeply until we were below the cloud base and able to pinpoint our position as we crossed the Lincolnshire coast. It would be distinctly unhealthy to cross the coast outside the authorised corridor and it would be almost equally unhealthy to descend through the cloud in total darkness without radio control or assistance, but in those days we had no choice. In 1940 the whole business of flying in Bomber Command was unhealthy and I never heard of anyone who wasted time pondering about the value of his pension rights.

'A surprise call came from the pilot. 'We're losing power, it's carburettor icing and I can't shift the Warm Air Lever. Come up and see if you can manage it'. I unplugged my helmet, then disconnected my oxygen tube (an acceptable risk at that height) and scrambled through the confoundedly small and obstructed passage with all possible haste to get to the space behind his armoured seat. I tugged at the control for all I was worth, well knowing that we faced real trouble. My efforts were fruitless; it was frozen solid and never moved. In a few moments the port engine had stopped and the other was giving less than half power. The nose pointed down and we approached an unfriendly North Sea at a rapid rate.

'I estimated that we would be ditching within seven or eight minutes and returned to my table to draft an SOS message complete with an estimated position. Our wireless operator started to transmit and his instructions were to repeat it continuously until we were about to hit the water. There will be no Christmas for us I thought as I tried to dismiss from my mind a store of unpleasant facts. None of us had any illusions; there had been too many reports of the disasters which had taken place when others had tried to do what we were now to attempt. The chances of getting safely into our rubber dinghy were far from rosy. Ditching an aircraft at sea was not a thing that pilots could rehearse. Frequently they stalled too high and crashed, thereby knocking out all the crew. Sometimes they nosed into a high wave and sank immediately.

'Of those who climbed successfully into our primitive type of dinghy, very few were known to live for more than two or three hours when cold, wet and exposed to the sky in winter weather. The Air Sea Rescue Service for the North Sea existed mainly on paper in the planning stages and minefields were so numerous that they formed a serious obstacle for rescue craft. In short, not many ditchings had a happy outcome. Air Force jargon had a phrase which fitted our situation. We had 'bought it'. I reflected that at least we did have a dinghy; in the past when there were shortages we had occasionally flown without one. Moreover our dinghy would not float away before we could board it, as a new instruction ensured that all dinghies were attached by a long length of stout cord. Pure chance had recently betrayed the unpleasant fact that many crews had seen their dinghies swept out of reach by wind and wave with fatal results. The lecturer from Group Headquarters had stressed that on no account was the cord to be removed or interfered with in any way. I had asked why such an injunction was thought to be necessary, what kind of fool would wish to meddle with it? 'Well... there is one born every minute... so they say', was the reply.

'Whilst all these thoughts were racing through my mind, our rapid descent was continuing, mercifully the remaining engine was still giving a little power which would greatly help us to make a satisfactory ditching. Three or four agonising minutes passed and then I heard that confident voice of 'Tubby' our rear gunner. 'I'll have a go' he said. I could not see what he was up to and had no faith that he would achieve anything, but he soon announced that he had 'shifted it.'

'There was no immediate improvement, one motor was absolutely cold so there could be no warm air reaching its carburettor, but the other one might well reap a benefit. But would it arrive in time? We broke through the low cloud base and got a chilling sight of a merciless black sea. This was the signal for us all to adopt our safe positions for ditching and I soon braced myself behind the pilot's armour-plated seat whilst the others sought refuge just behind the giant metal box-spar. The sea grew nearer and the beam of our landing light showed up the choppy water. Suddenly the pilot shouted that the engine was picking up a little and after a few more seconds we heard the sound of a perfectly healthy power unit. The Hampden bomber had one outstanding virtue in that it could maintain height on one motor and we found ourselves skimming the wave tops in level flight. We had escaped by about 30 seconds.

'I'll keep trying the other engine' yelled the skipper. Some minutes elapsed before the carburettor ice surrendered to an outside air temperature that was now relatively warm; there were coughs and splutters and finally a full roar. The crisis was over, we had two engines again. Within about ten minutes Skegness pier obligingly appeared in front of us and it was plain sailing back to base. Soon we were all tucking into the famous operational breakfast of bacon and eggs. I found time to stare across the table at 'Tubby' and to ask, 'How did you manage to move the Warm Air Lever?' 'Well it was like this, ' came the answer, 'I thought a bit of rope or something like that might help to give the necessary purchase, so I tied a length of cord around the lever, put both my feet against the main spar and pulled like hell'. I gazed in admiration at the man who had saved our lives. After a little thought I asked a second question: 'Where did you get the cord from?' 'Tubby' smiled, leaned forward and whispered, 'I took it off the dinghy. There's one born every minute... Remember? [64]

'The tempo of World War 2 gradually increased and life for everybody became more dangerous and inconvenient. In the Royal Air Force we naturally tried to meet each and every difficulty with traditional good humour, but there was one comparatively minor item which caused long faces. That was the lack of hot water for shaving; unsightly cuts and abrasions were testimony of lost battles with brutal razors. One memorable morning I entered the barracks to be confronted with an astonishing spectacle, a newly arrived airman stood with a mirror propped against the window frame, his hand grasped something that was connected to the light socket by a flex and a weird sound filled the air. The stranger commenced to pass that noisy object up and down his cheeks and around his chin causing the dark growth of beard to disappear as if by magic.

'It would have been naive to display open amazement, one should never allow a recruit to think he has created an impression on an older hand in the Service as that might impair his sense of respect. I indulgently refrained from sternly announcing that meddling with electric light fittings was a contravention of the Station Standing Orders so prominently displayed only a few feet away and I adopted a contrived casual approach which would have delighted those old friends who had striven to give me a little training in amateur dramatics some years previously.

'What is that you have there?' I gently enquired. 'This? Why this is an electric razor... the very latest thing to arrive from the States, believe me they are absolutely marvellous'. I suppressed my intense admiration and betrayed no sign of special interest, but it was certainly a case of love at first sight. The electric razor was the perfect answer to my daily difficulty, we never lacked for electricity and the magnificent invention was an urgent necessity. The afternoon found me scouring the shopping centres and eventually a proprietor in the High Street gave me the right reply. 'You are lucky chum, we have never had many in stock and this is the last one. Under the new import restrictions there will be no more until the war is over and who can tell when that will be?' From that moment life was happier.

'Promotion came at a later date and I gained a private room as a privilege, this enabled me to shave in comfortable seclusion and I revelled in the luxury. Before long I was posted to a Bomber Command Training Unit in dear old Lincolnshire and my new comrades were left in complete ignorance of the treasure in my possession. This elite organisation was under the command of a veteran old fashioned autocrat who, without any conscious effort, projected the most glamorous image of a born leader, the sort of man that scores of professional post war actors have vainly tried to portray in films and on television. Our much respected Station Commander would never concede that a war provided excuses and on our daily parades all ranks were expected to match the turnout of the Brigade of Guards from which he himself had transferred in 1916. In my new role of Chief Ground Instructor I was prominent at every ceremony, trying faithfully to provide a shining example. Then came a terrible day when I was grooming myself and found that the very worst thing had happened, my electric razor would not work because we had a power cut.

'It seemed to me that it was just possible that the disaster might not be universal in all areas of our widely dispersed camp and I hastily drove a mile around the aerodrome perimeter to my office overlooking the parade ground. Upon arrival I was shaken to find that the Group Captain was already there, resplendent in all his usual magnificence. How I cursed my ill fortune, why on earth should he have chosen this particular day to arrive so early? Evasion was impossible so I saluted formally and brazenly said 'Good Morning Sir'. He stared unbelievingly, his eyes narrowing until they formed mere slits, I could almost feel the impact on my conspicuous black stubble and braced myself for a audience, he turned abruptly away in the direction of Station Headquarters which lay at a distance of three or four hundred yards. His intention was crystal clear; he would telephone my office and demand my instant presence so that I could receive one of the most devastating rockets of all time.

'My career was in jeopardy. I had apparently completed two tours of hazardous missions against a formidable enemy only to be felled on my home ground. It was common knowledge that a Nazi officer who lapsed from grace could expect to find a revolver with one live round by his bedside. In all probability our Group Captain would not go as far as that but he might well decide that he had had quite enough and send me off for a third tour with the Bomber Force where the survival rate was phenomenally low. I charged into my office and found that the light switch worked, then I leapt on to the table with an agility born of desperation and snatched at the light bulb which dropped on the floor with a crack like a gunshot. In went my cherished razor and within about one minute every trace of offensive beard had gone. Almost immediately the expected phone call brought me a firm order and I was soon gently knocking at his office door. A very loud voice shouted 'COME IN!' The Group Captain sat at an imposing desk, his handsome head face downwards seemingly in close contemplation of some important papers which he shuffled in a restless manner. After a suitable pause he threw them down, banged both fists on the desk and barked 'NOW...' That is about as far as he actually got, the face that was usually so controlled and dignified became a sad picture of bewilderment. His eyes nearly popped out of their sockets. I gazed back calmly. 'Yes Sir? You wished to see me?' His jaw moved up and down in a fashion reminiscent of a goldfish, with a contorted smile he stammered a request for my opinion about the weather prospects for the flying programme. Stepping briskly up to the large window I contemplated the sky in the full knowledge that the morning sunshine would illumine my face as I thoughtfully stroked at my very smooth chin. Later on at our customary tea break our Adjutant commented 'I must say that it is very unusual for the Station Commander to take a morning off at short notice. I heard him say on the phone to somebody or other that he did not feel too well. [65]

Haworth eventually departed in December 1944 having flown two tours of operations and two tours as an instructor. 'By that time' he recalls 'I had seen and learned a lot. Most of the aircrews trained in the thirties were wiped out completely by that time. Guy Gibson's entry was such an example. The question of crew casualties is a vexed one; the true rate never comes to the surface at all. There were 25 men on my course in 1938. Three years later I was the only survivor. The truth is that Bomber aircrews fell like the leaves in autumn. The burden borne by the early participants is in no way brought out. In and around 1941 and 1942 the casualty rates at both Hampden OTUs were very high. I have no evidence that those figures were included in Bomber Command statistics. They were considered to be merely under training. I was told that special measures were initiated to ensure that the output from Cranwell College did not entirely vanish. The position was skilfully played down by the Press and wireless. Under a policy originated by the great Lord Trenchard, by 8am each day, crews lost during the night had been replaced, there were never visible gaps in the ranks. Everybody was very occupied, all aircrew tended to look alike. The changing faces attracted no particular comment. I well remember that originally, when returning from leave or some brief absence, I would ask 'Has anyone been killed?' After a short period of time I started to ask, 'Who has been killed? [66]

In May 1941 Sergeant Frank Lowe, a 21 year old Hampden pilot on 49 Squadron at Scampton in Lincolnshire set out on his fourth raid, on Hamburg docks. 'It was a moonlit night and we made the 400 mile sea-crossing without incident. We crossed the enemy coast a few miles north of the Western end of the Kiel Canal, successfully avoiding a concentration of searchlights, then turned onto a South-Easterly heading until we were East of Hamburg. Then we turned to starboard and made our bombing run across the city to the target. The flak was intense and the searchlights persistent but the docks showed up well in the moonlight and we were confident that our bombing, from about 9,000 feet had been accurate. After leaving the target area we started a steady descent, intending to make the return trip at low level over the sea, thus reducing the risk of being spotted by night-fighters. Unfortunately the sea by this time was covered by a vast blanket of fog, brilliant white in the light of the moon. Our Hampden would show up clearly in silhouette to any enemy aircraft above us. I chose the obvious solution and gratefully let the aircraft sink into the welcome concealment of the fog until the altimeter read 400 feet. I dared not go any lower as I did not know the local barometric pressure. We had all relaxed and were already savouring the prospect of eggs and bacon for breakfast when all hell was let loose-we had flown straight over a flak-ship! We were surrounded by spirals of tracer - a mixture of machine gun and 20mm cannon fire, the latter almost certainly quadruple-mounted. I instinctively flew an evasive corkscrew. My under-gunner asked me whether he should open fire and I hastily told him not to, as it would have given away our position. After what could not have been more than half a minute, but seemed much longer, we were out of range and unscathed, feeling as much indignant as frightened! We really enjoyed our breakfast safely back at Scampton!

'We were briefed for a bombing trip to Bremen, despite a poor weather forecast. It was a dark, wet February evening when we got airborne in our Hampden, penetrated the low cloud-base set course for the target and slowly climbed to 6,000 feet still in cloud. As we neared our ETA I started a slow descent and we eventually saw the glow of searchlights, hopefully indicating the target area. We finally broke through the cloud-base at about 4,000 feet. The visibility was poor, with misty drizzle, so I told my WOp/AG to drop a flare in the hope of identifying the aiming point-the docks. By this time we were down to about 3,500 feet and I suddenly became aware of a brilliant light below us, lots of smoke in the cockpit and a strong smell of fireworks! My WOp/AG called me to report that the flare had stuck in the flare-chute and ignited. I told him to try and knock it out with the fire-axe. Needless to say the German gunners and searchlight crews had by this time chosen us as a prime target. We had some very near misses from light and heavy flak and were coned by searchlights. I took the usual evasive action, opened the bomb-doors, told the navigator to drop the bombs, opened the throttles and climbed back into cloud cover. My WOp/AG had managed to dislodge the flare and we thankfully set course for home.

'One summer day we were told that we would be 'on ops' that night and had air-tested our aircraft that morning. In the afternoon we had attended the briefing. We had an early dinner in the Sergeants Mess - soup, main course and sweet - and then an hour or so of rest before walking to the 'flights' to don our flying gear before being trucked out to the aircraft in dispersal. It was standard practice for the whole crew of four to gather round the tail of the aircraft to spend a last penny (pee now or forever hold your pees!) before climbing aboard. On this occasion the under-gunner found it necessary to move a bit further away so that he could dispose of something more substantial than a penny! We should have been alerted by this non-standard behaviour. However we finally took our places in the aircraft, blissfully ignorant of the unpleasant ordeal we were soon to endure as an extra to the hostile actions of the enemy. We did our initial checks, started the engines, went through the after start-up routine, taxied out towards the take-off point, carried out the pre-take-off checks got our green light and took off. As far as I can recall, the North Sea crossing was uneventful, although always potentially dangerous. Unless there was plenty of cloud cover we always felt very vulnerable there, knowing that we presented a silhouette against the light of the Northern sky to any enemy night-fighter approaching from the dark of the South. This feeling of vulnerability probably produced an occasional spasm of the anal sphincter but on this trip all four of us were experiencing the symptoms of incipient diarrhoea by the time we reached the enemy coast. We had to exercise extreme muscle control to avoid personal disaster! To make matters worse we found that the cloud base over Emden, our target, was only about 4,000 feet.

'We did our bombing run at 3,000 feet, nearly got coned by searchlights and survived some very near misses by light and heavy flak. On the way back home, we seriously considered the possibility of landing in a field in Holland, so that we could relieve ourselves! This was not on of course but we did keep our speed up to 150mph instead of the usual 130mph for a lightly loaded Hampden. We eventually landed at Scampton, taxied as fast as possible to the hangar, stopped, switched off and we were all out and running for the latrines, discarding flying gear as we went! I nearly made it but had to waddle awkwardly the last few yards. I heard my under-gunner scramble into the next cubicle - he was actually dropping his trousers when the worst happened. The other two members of my crew also lost the race! We heard later that another of our pilots, upon suffering the symptoms on the outward trip, had made a brave and imaginative effort to deal with the problem. He engaged the autopilot and told his under-gunner to come up behind the pilot's seat with the paper bag which had contained the 'rations' - chocolate etc. The Hampden was a very narrow aeroplane - no room to sit side by side - and the back of the pilot's seat was hinged so that, in an emergency, it was possible to change pilots in flight, albeit with some difficulty. This chap undid his Sutton harness, dropped the seat back, took off his parachute, eased his body out of his Sidcot flying suit, pushed down his trousers and pants and got his gunner to hold the ration-bag where the toilet bowl should have been. Having relieved himself, he went through the rigmarole of pulling up his clothes, re-donning his flying suit, putting on his parachute pulling up the seat back and re-securing the safety-harness. Unfortunately his efforts were in vain because he had an uncontrollable spasm about ten minutes later! The gunner took the ration-bag with its repulsive contents back to his cupola and later dropped it over the target area. There was, of course, quite a stink about this episode, in more ways than one.

'We were issued with new clothing. An investigation by the MO put the blame on the soup served in the Sergeant's Mess. None of the officers was affected. The cook responsible was quietly posted to another unsuspecting station!

'I can't remember the exact date but I believe it happened in March 1941. We were returning from another raid on Hamburg, crossing the North Sea and slowly losing height. The half moon was low in the sky and there was scattered cloud. About halfway across we were beginning to relax when a burst of tracer zipped past us, too close for comfort. I took violent evasive action altered course and knocked off some more height. No longer relaxing we resumed our course for home. About half an hour later we were attacked again in an almost identical episode! Once again we managed to shake off the enemy aircraft. I told my wireless operator to send a message to our base at Scampton to warn them that enemy intruders were approaching the coast on our heading. In due course we crossed the English coast and turned onto our final course for Scampton. With the city of Lincoln in sight I suddenly spotted a Ju 88 400 yards ahead of us. I warned the crew and opened the throttles in an attempt to close up on him and give my gunners a chance at him but he was travelling too fast. We joined the circuit at Scampton, flashed our identification letters got, an answering green from the Aldis and landed. We were taxiing to our dispersal when a shower of incendiaries fell across the field. We saw them bouncing into the air as they ignited upon hitting the ground. Then we saw a burst of tracer in the sky as a Ju 88 shot down a Hampden in the circuit. The pilot and wireless operator of this Hampden managed to bail out but the other two crew members were killed. The pilot had a lucky escape - he saw tracer passing between his legs! This experience finished his flying career - he took to the bottle!'

The sole survivor in the UK in 1985 of the very first aircraft in, on the very first Combined Operation, recalled: 'About ten selected crews of 50 Squadron, in early December 1941 were transferred to Wick, where their Hampdens were flown on many training flights. Navigating to pinpoint rocks in the Atlantic, liaising with the Royal Navy, low level and formation flying, were practised daily. Skills improved except for night formation flying which was nearly impossible and was thus discarded. So even in training, lessons were being learnt. Christmas morning brought 'Ops On', so the Night Flying Tests were all completed by noon, just in time for us all to serve out the traditional lunch and festivities in the airmen's mess. The Officers' and Sergeants' Mess exchanges were dry and of short duration, as the afternoon brought a lengthy briefing session. The raid on Vaagsö and Maaloy was explained in full at first and then in detail as to the role of each crew, our part being to go in by a fjord to the south, penetrating further inland to Ragsundo where three large guns were positioned. As these guns pointed down the fjord threatening the Navy ships and the Army assault craft approaching Vaagsö, it was our job to put them out of action. Christmas evening in the hut where the 50 Squadron senior NCOs were temporarily billeted was quiet until the CO, Wing Commander Oxley and a few officers walked introducing a bucketful of beer, he invited everyone to have a Christmas drink on him, which brought so many familiar answers, 'No thank you sir, I'm flying tonight'. Only then did he divulge that as HMS *Kenya* had screw (propeller) trouble, the operation had been postponed 24 hours! That bucket was emptied in seconds, followed by feverish activity as everyone prepared for the Sergeants' Mess Dance. It was a memorable night as 50 Squadron Officers and NCOs led everyone in the festivities.

'As dawn was breaking on 27 December, we were spot on at low level, flying up the fjord. As we caught sight of the German billet, the door opened and soldiers tumbled out, donning their hastily grabbed overcoats as they ran along the path towards the guns. The under gunner and I had a great opportunity to empty our pans of ammo at the scampering 'bods'. Seeing and overflying the gun site, we swung round the comer of another fjord where we climbed to do a 180 turn on to our bombing run. Just as the bombs were released a cannon shell hit the oval aiming window in the nose of the aircraft. John Day the navigator was, of course, right over the bombsight and took the full blast of the explosion. Nevertheless, he called out over the intercom: 'Bombs gone. I've been hit. The course for home is 230'. He then pressed his seat button and as it swung back, he fell to the floor since his legs were badly damaged. Our dash for the nearest landfall at Sumburgh in the Shetland Isles certainly saved his life, but not his leg, as that had to be amputated in Lerwick hospital. He regained health and strength and, with an artificial limb, got back to flying again in the Mosquitoes of the Light Night Striking Force with Pathfinders.

'One of the other 50 Squadron Hampdens which was dropping smoke canisters on to the beach, to screen the approaching assault craft, was badly hit. The severe damage to the wing made the aircraft swing out over the water, losing valuable height all the time. The pilot, Sergeant Reginald N. Smith of Wankie, Southern Rhodesia was a big strong lad and he managed to gain control enough to make a good splash down, but the Hampden being a notoriously poor 'floater' sank very quickly. The navigator, pilot and wireless op were all clear, but the rear gunner, Sergeant Derek Bell was just climbing out of the hatch when the kite went down, taking him with it. The hit on the wing must have damaged the dinghy since it never appeared. However the Mae Wests kept the surviving three crewmen afloat and they grouped together. Smithy was the only able swimmer so the navigator and wireless operator held on to the trailing mangle of his Mae West, as he struck out for the shore. Although land was only about 50 yards away, the strong current, the intense cold plus the extra load of his two crew mates, prevented Smithy making much headway.

Meantime, all three men were shouting for help. Round the bend of the fjord and out of sight, someone on HMS *Kenya* thought they heard shouts. A call for quiet confirmed this and a whaler was immediately sent to the rescue. It picked up all three men but, by the time they reached the *Kenya*'s Sick Bay, both the navigator and wireless operator were found to be dead from exposure and cold, although they had been in the water for approximately 15 minutes. Through the energy and exercise of swimming in trying so desperately to save his crewmates, Smithy's circulation had kept going and this inner warmth had saved him from a similar fate.

Smithy was soon back to his normal self and during his days aboard the *Kenya* on its way back to England, he was approached by one of the officers, who explained how lucky it was that someone aboard Kenya had managed to hear the shouting. This officer presented Smithy with a whistle, saying that it took a lot less energy to blow a whistle than to shout, that it produced a louder and distinctive sound which carried much further and could even be used to send messages by Morse.

'It was into January 1942 when HMS *Kenya* got back to Britain and Smithy was de-briefed. Only a few weeks later on the 7th February, a few minutes after noon Smithy took off from Skellingthorpe for a daylight mine-laying operation off the Frisian Isles, but failed to return. I just don't know if Sergeant Reginald N. Smith had lived long enough to know that his experience had not only been another good lesson learnt through Combined Operations, but also brought about the issue of whistles to all aircrew thereafter.' [67]

Endnotes Chapter 10

64   *One Night In December* by Gilbert Haworth, Lincolnshire Life, December 1983.

65   *My Closest Shave; A true narrative* by Squadron Leader G. Haworth DFC DFM writing in Intercom; the Official Quarterly Magazine of the Aircrew Association, Christmas 1985.

66   'Total Casualties in Bomber Command's aircrew were: 51% killed on operations against the enemy. 9% killed in crashes in England. 3% seriously injured in crashes. 12% became prisoners of war - some of these were injured. 1% shot down but evaded capture. The bulk of the 24% who survived unharmed served in the last few months of the war. I paid a visit to Bomber Command HQ in July 1943, visited the Personnel Section where I learned that up to that time only 8% of bomber crews had ever successfully completed one tour and that survivors of a second tour were so few that a useful graph could not be compiled. I did a little research on 207 Squadron covering the first six months of 1943. Some aircrews did not last a week; the average stay was 30 days. A fat lot of leave they got. Most were actually killed.' Gilbert Haworth, writing in *Intercom; the Official Quarterly Magazine of the Aircrew Association*, summer 1995.

67   *Intercom; the Official Quarterly Magazine of the Aircrew Association*, Christmas 1985.

Chapter 11

Wimpy Ops

It is just a few nights ago that I am making my first mission with the Royal Air Force against the enemy. I was for four years observer in the Czechoslovakian Air Force. I go to France and I am nine months in France in the French war. Then I come to England with a little ship from Bordeaux. It is a little merchant ship. Eight hundred tons. A little Dutch ship. I am in some, camps here and then I go to a Czech depot. I do not speak any English when I arrive. I buy an English dictionary and when I have free time, I learn to speak it: but it is not so quickly. Then, since September, I am in the Czech bomber squadron with the RAF. I have a course of navigation which is most painstaking and in October I am ready, but I must wait for my pilot. He is not so ready. We are to fly Wellington bombers. It is very good, the Wellington.

So, I am waiting to go for some weeks, but my captain is not yet fully instructed. Then, when he is ready and we are to go he becomes ill. He has influenza and after more waiting I go on this, my first mission, with another crew. Before this happens I have asked already to go with another crew until my pilot is finally prepared, but this is not possible, for all the other crews are complete and it is not good to chop and change crews. But now I can go. We are on this occasion to bomb the submarine base at Lorient. Our pilot is a sergeant and we have altogether three other sergeants and two officers. Before we go we are briefed most exactly by our own Czech wing commander. The wing-commander tells me that I must find exactly this target to bomb. He gives me the position of anti-aircraft guns and the position of submarines in these docks at Lorient. When we are going out the weather is good. It changes very quickly, the English weather: now good and then after two or three hours it is, perhaps, bad. This was good all the time except that when we come back we have some clouds-but that is not yet.

When I come to the French coast I am finding myself a little off my course. You see, I am to be here at a certain point and I am four miles from this but I fix me by the coast and we go on. The aircraft goes very well indeed. Before we have come to Lorient we have seen a big fire: I think about twenty-five miles away. Also big anti-aircraft fire and searchlights - many of searchlights - and much of anti-aircraft fire. When I see all this I say, 'That is Lorient where we are to go', but we are already going in this direction so it is not necessary to make an alteration. We go into the anti-aircraft fire and into the searchlights. They fire at us but the explosions are not of our height. The others with me have already made eight or nine missions with our squadron and have been under fire very often, but for me it is the first time. It looks very nice, I think and illuminates the night very nicely.

We are flying into our target from the north. I go a little to east and turn also back to south. Before this, I have prepared my bomb-sight. Then I am looking and giving the directions for my pilot. I have seen the river and on the right side of the river I have seen here a big building which must be the smithery which is shown for me on my map. I have seen the river very good. It looks a little lighter than the ground and it is quite simple to see. Also I can see the docks. There is no cloud at Lorient: only little patches of mist, but we can see through them quite simply. When we get near we see again the fire that we have seen before, but now he was not so great: he is dying. Since I am to make the bomb-aiming I am lying on my stomach in the nose of the aircraft looking down on all this. I have, besides heavy bombs, many incendiaries. When I have the target in my bomb-sight I tell to the pilot 'Straight on'. He goes straight on and when I have got this target as I want it I press the switch. I can see the bombs go from the aircraft. Just for one moment. No more. The rear gunner says he has seen the explosions from the bombs. Then it is most remarkable what happens. There are between twenty and thirty big explosions and soon they are making altogether one big fire. There is much talking in the aircraft at all this and everybody asks the others to look at what is happening. The front gunner is pleased and cries away that it is the biggest fire he has seen in all the nine missions he has made. All this time, they are firing at us also heavily, but again it is a little down and not of our height and we are not hit.

It is very good because this fire had held on so well that when we go back from the target twenty minutes the rear gunner is finding it still possible to see the fire. When we are going back, there is more fire at us for one or two minutes, but still we are not hit and we come home without difficulty. When we are home once again the wing commander is very pleased with what we have done and tells us that this was a very good show.

Raid On Lorient by a Czech pilot.

Affectionately known as the 'Wimpy' after the American cartoon character J. Wellington Wimpy in 'Popeye' who was always eating hamburgers and said, 'I'll gladly pay you Tuesday', the Wellington had been designed by R. K. Pierson using the unique method of construction called geodetic or lattice work structure conceived by the brilliant British scientist, Dr. Barnes Wallis. The Wellington and the Hampden and the Whitley were weakly armed but it was these bomber's exploits, which featured in the headlines in the British press and sometimes in German papers as well. In a contemporary account Sergeant Myers, rear gunner on Pilot Officer Reynolds' Wellington IC R1088 'O-Orange' crew on 101 Squadron at Oakington, Cambridgeshire, recalled the trip to Hamburg on 30/31 August 1941.

'I remember very well,' he says, 'going to the Intelligence Room that afternoon and the feeling of expectancy I had when I learnt that my target was to be Hamburg. I knew that new fighter patrols had recently been seen round there-they were marked, with the usual red wool, on the map in the Intelligence Room - and I suppose that is why I felt as though something exciting was going to happen. On the way out we were all on the alert and everyone of the crew who was available carried out what we call a 'search' of the sky very thoroughly indeed. From my rear turret I searched continuously and I certainly felt relieved when we got near Hamburg without meeting anything except occasional searchlights and flak. We got away from them by evasive action and were able to continue on our course without undue delay. We got to Hamburg well in time. As far as I can remember we were not satisfied with our first run over the target, so we flew over the area for about twenty minutes. The flak, both heavy and light, was particularly intense, but eventually we got a good run over the target and saw our bombs burst on it.

'We began our long journey home and things seemed to be going particularly well, when we suddenly began to get into trouble with searchlights. The pilot took evasive action, but in spite of everything he could do we were held by more and more searchlights, until in the end we found ourselves in a cone of about thirty with very intense flak all round us. The searchlight crews appeared to be working on a definite system; their idea seemed to be to hold us in the centre of a huge circle of light with a diameter of about 400 yards. We were feeling, as always when caught in searchlights, as though we were completely naked, when the flak suddenly and entirely stopped.

'I knew what this meant at once and I did not need the captain's caution, 'Watch out for fighters.' To anyone who has not been through just that moment of suspense it is difficult to describe the eerie sensation one feels. It is as if time were standing still and the whole world waiting. I seemed to be in the centre of an enormous lighted but empty circus, waiting for the unknown. Obviously I knew in a sense what I was waiting for, but I felt that everything was fantastic and unreal.

'I kept my eyes skinned and tried to look in all directions at once. Suddenly, I caught sight of a faint ghostly shape entering the circle of light. At the same time streaks of light were slithering under my turret. Of course I realized that they were tracer, but my brain refused to register the fact that they were deadly. I immediately swung round to face the attacker. But the attack was so swift that it was almost over before I could bring fire to bear and I fear that my sighting was in consequence rather sketchy.

'I told my captain of the attack and then in quick succession there were four more, each by a Me 110. My captain thought afterwards that the last attack was a repeat by the first Messerschmitt, but my own opinion is that the attacks followed one another too quickly for that to have happened. After my rather erratic burst at the first fighter I was ready for the others. I fired between eight hundred and one thousand rounds at them at point-blank range; at least half my shots must have found their mark. Not a single shot from the fighters hit my turret; all their shots went by underneath me, but a great many bullets and one or two cannon shells must have struck the starboard wing and the other side of the fuselage.

'Three of the fighters never appeared again, but I could not claim them as victims of my fire because there were no positive signs of damage to them. The other two fighters each made a second attack, one from the beam, to no effect and the other from the front. In this attack a cannon shell wounded the front gunner; it passed through his left arm and went on to smash the pilot's instrument panel; only the compass and the altimeter were left working. The Wellington went into a very steep dive. We hurtled down from 10,000 to 2,000 feet before we pulled out. I could get no answer from the intercom during the dive and you can imagine my feelings; I was on the point of bailing out when we began to pull out.

'Our dive shook off the fighters and at the same time we somehow got away from the searchlights. But the engines were, missing badly. The captain sent the second pilot back to see if I was all right and to warn me to be ready to bail out if the engines failed altogether. The others with difficulty took the front gunner from the front turret and brought him to the bed to the rear of the main spar. They applied a tourniquet to his upper arm and did everything possible for him.

'The engines unexpectedly picked up again and we were able to gain height gradually, though we found to our consternation that we had lost three hundred gallons of petrol from our starboard tanks. The pilot consulted with the navigator and decided to try to make for the Dutch coast before baling out; he hoped that we might then be able to escape to England by boat. But by excellent airmanship and skilful saving of petrol, the pilot discovered that at to the Dutch coast there was enough petrol to take us within sixty miles of the English coast. We decided to take a chance and if necessary, even though our wireless had failed, come down in the sea. Actually we flew for half an hour after the petrol gauge had registered zero and with great relief we eventually crossed the English coast. The navigator had brought us direct to his aiming-point, an aerodrome near our own and just inland. Exactly as we were gliding in to land our engines cut, but the pilot made a wonderful belly landing. Our wounded front gunner, after this struggle of three and a half hours to get back to England, died in hospital, half an hour after he was taken there.' The front gunner's name was Pilot Officer David Lecky Crichton RAAF. Another member of the crew, Sergeant Owen Hugh Barlow, was killed. The Wellington crashed at Rampton, seven miles NNW of Cambridge.

'To such defences' said a contemporary report 'there are two main answers, tactical ingenuity and the fighting spirit of bomber crews. The first answer is the strong and thorough construction of all types of bombers in service, the workmanship which enables them to get home with one engine out of action, with undercarriage unusable, with instruments broken, or with a hundred other kinds of damage from anti-aircraft fire or the guns of fighters. After bombing by daylight, a Wellington was attacked by a German fighter. Bullets whizzed between the Wellington pilot's legs and went out through his cabin; others just missed the second pilot. An oil pipe in the rear turret was cut, the rear gunner was wounded and his whole turret was soon enveloped in flames. The gunner put them out, scrambled back to his seat and found that he could hardly see from his turret because the perspex was covered with oil. But even so he was able to fire back at the fighter when it came in to make another attack. In this second attack the gunner was wounded again and again his turret caught fire. The second pilot came to his help and he was able to crawl back to the fuselage. But the fire spread to this also and the bomb doors were set alight by incendiary bullets. Second pilot and wireless operator fought the fire with everything they could find, extinguishers, gloves and cushions; they threw burning material overboard.

Suddenly both engines stopped and the Wellington went into a dive. The captain was thrown forward and a hook of his parachute harness caught on the control column. This left him quite unable to control the Wellington while it was going down from 10,000 feet. The navigator came to his help, but as he was trying to get the hook undone a bullet struck him. The captain had to struggle by himself to get free, but did not succeed until the Wellington had fallen 8,000 feet. Then he was able to flatten out and, by extraordinary good fortune, both engines came on, one after the other. When the time came to land, the pilot found that his undercarriage was so damaged that it would not go down. He asked his crew whether they would rather bale out or risk a landing. They said: 'No. We'll stick with you, sir.' He went on and made a crash landing without any further injury to his crew. They looked at their Wellington after the landing and saw that it was almost a skeleton; so much of the fabric had been burnt away. There could have been no more complete test of workmanship and design.

Equally tough was the bomber which was brought home from Germany after a collision with a Messerschmitt. The Messerschmitt had intercepted the bomber soon after it had turned for home and in the first exchange of fire one of the bomber's engines was put out of action. Bullets tore through the fuselage, smashed the wireless set and wounded one of the crew in the shoulder. The Messerschmitt then came in to attack from the rear, but the pilot was too daring and the tip of his aircraft caught the bomber's rudder and took half of it away. But the Messerschmitt had the worst of the collision and was last seen diving steeply away, apparently out of control. The pilot had now to use all his skill and strength to control the bomber; he managed to keep it on a fairly steady course, but he could not climb higher than 700 feet and for mile after mile he hobbled along towards the Dutch coast. Towns and villages looked uncomfortably close and as the pilot said, 'If the Germans were watching they must have thought that we should crash at any moment.' As the bomber crossed the coast the pilot told the wireless operator to crawl down the catwalk and let the rear gunner know that as the intercom had been damaged in the fight a torch would be flashed if they had to come down in the sea; the rear gunner could then be ready with the rest of them to get into the rubber dinghy. Everything that was loose was thrown overboard to lighten the aircraft. But this was not enough and no wonder, for even as they were throwing everything out they found that a heavy bomb was still in the bomb rack; shrapnel from the anti-aircraft guns had no doubt broken a part of the release apparatus when the bomber was over the target. To get at the bomb and jettison it the crew had to hack away some of the fuselage; then the pilot was able to climb to 1,000 feet. The aircraft was still difficult to steer, but the pilot stuck to his course, crossed the North Sea and landed without any further mishap on the first aerodrome near the coast.

On the night of 7/8 September 1941 when 197 aircraft attacked Berlin and another 51 raided Kiel, a great number of fighters, not far short of a hundred, were seen over Germany and the invaded countries. A Wellington was held in a cone of searchlights over Holland; suspiciously, there was no anti-aircraft fire and the suspicion was confirmed when the rear gunner suddenly saw tracer bullets coming from 300 feet above and astern. He could not see the enemy fighter at all, but he fired back, three bursts in all, in the direction from which the tracer was coming. No more tracer came after the first burst, but instead, the gunner saw small objects, black and smoking, fall down through the searchlight beams. Evidently the enemy fighter had been severely damaged, but nothing was ever seen of it, though its defeat must have been seen from below, for the anti-aircraft guns on the ground at once began to put up a fierce barrage. At other times the presence of enemy fighters was only detected when they fired coloured lights as a signal. One such fighter got itself into a barrage, signalled to the gunners to cease fire, but was hit and fell in flames.

Over the northern outskirts of Berlin a Hampden was held by searchlights and dived to 4,000 feet to get away. At that height a fighter approached and fired from 100 feet away. The Hampden pilot went on with the dive until he was within 50 feet of the ground, but even at that height the bomber was still held by the searchlights and it was not until it had flown several miles just over trees and roof-tops that it got away from the lights and from the pursuing fighter. Sometimes the enemy appears to have been less persistent. Over Holland a Junkers 88 circled round a Wellington and then at once made off; a Stirling turned to attack a Messerschmitt 109 and the enemy at once turned and fled. Over western Germany a bomber was being engaged by anti-aircraft fire; a Junkers 88 approached, did not like the look of the barrage and would not come up to the attack.

A Junkers 88 attacked a Wellington four times over Berlin and got in one or two hits. But when it made its fourth attack the Junkers flew right through a stream of bullets pouring from the Wellington's rear turret and as soon as it was through the zone of fire it caught alight and fell to the ground. The Wellington crew were able to see the wreckage burning on the ground for some time. In all, four enemy fighters were claimed destroyed that night and several more were damaged. Fifteen aircraft were lost on Berlin; two Hampdens and a Wellington failed to return from Kiel.

Though it was not until after the spring of 1941 that night fighters became most active over Germany, at no time could any gunner of Bomber Command afford to relax attention for a moment. Every battle is an unrepeatable event and no one story can do duty for the sum of the experience of all the gunners of Bomber Command. But the story of a sergeant rear gunner - I have given him another name than his own - does at any rate show what these men must be ready for. It happened on a night in February 1941; the gunner was in a Whitley, on its way to Bremen.

In 1941 there were new bombs as well as new bombers. These were larger than any previously used by the RAF before; they weighed 4,000 pounds each. They had no particular connexion with the introduction of heavy bombers, for they could be carried by medium bombers as well. They were first used on the night of 31 March / 1 April in an attack on Emden when six Wellingtons on 149 Squadron at Mildenhall were dispatched and 28 Wellingtons attacked Bremen.

At Mildenhall on 3 March 149 Squadron had received two of the new Wellington Mk.II bombers which, with their powerful Rolls-Royce Merlin engines, were capable of carrying a 4,000lb high-capacity (HC) light-case bomb commonly called a 'Cookie' that was being brought into service. This bomb somewhat resembled a giant dustbin. On the night of 31 March it was first dropped in anger when two aircraft on 149 Squadron acted as cover for the Mk.II Wellingtons carrying the 'Cookies'. 'X-X-ray' piloted by Pilot Officer John Henry Franks[68] successfully completed the operation but the second failed to get airborne and slid to a halt in a barley field at the edge of Mr. Norman's small-holding at West Row - the *Wizard of Oz* painted on the bomber had failed to work its magic. The Wellington flown by Sergeant G. J. P. Morhen landed heavily on return, stalled and crashed. One gunner died of his injuries later. A second 'Cookie' was dropped by a 9 Squadron Wellington. Cookies had no ballistic characteristics. Once released, they could land anywhere within a five-mile radius of the aiming point. One of the Cookies fell in the east part of the town near the Post office and telephone Exchange, causing severe dislocation to these services. The other fell in the old part of the town. When it exploded, 'masses of debris' said the official communiqué, 'flying through the air were outlined against the glow of fires and the results appeared to be devastating.' 'Houses took to the air' said the pilot who dropped it. These 'high capacity' bombs soon came to be called 'Luftminen' by the Germans.

The pilot and navigator of the bomber which dropped the first of these bombs took particular care to aim well with it. The pilot's first run over the town, with fires below as a guide, was made from south to north. There was an intense barrage and the pilot had to take evasive action and so lost height, which meant that the navigator had to readjust his bomb-sight. When he had done this the bomber was almost directly above the target and it was too late to release the bombs then. The pilot turned and made a second run from the south-west; the navigator had everything ready now and the pilot flew straight and level through the barrage. A fire in the area of the target below made an excellent aiming-point as the bomb aimer saw it come along the drift wires of his bomb-sight. Even before the navigator had said 'Bombs gone' the rest of the crew knew when the big bomb went; it felt as though the aircraft had shaken itself free of the bombs with a distinct jerk upwards. They waited about twenty-five seconds and then there was a flash which lit up the inside of the bomber, flying at 10,000 feet. Immediately after the detonation the crew saw great masses of debris flung high into the air; from the height they were flying they would not have seen much of the debris thrown up by an ordinary bomb and they knew that enormous destruction must have been done. The debris must have been thrown to a great height because there was a distinct interval before it settled down. This bomb fell about 200 yards to the west of a fire; when it burst, the smoke and dust spread out over the flames and blanketed them from the watchers above. All they could see was a glow underneath and then in a minute or so the flames shot up again.

A very small force attacked Emden that night, but the Germans admitted in their official communiqué that 'comparatively large damage to buildings' had been done. This would have been a sufficiently unusual admission even if a very large number of bombers had attacked Emden in the ordinary way and photographic reconnaissance after the attack showed why the admission had been made. The areas of complete devastation where the bombs had fallen was very large indeed and by further reconnaissance, after the necessary work of demolition had been done, the efficiency of the new bomb was unmistakably proved. There was complete destruction over an area of about 2½ acres and blast damage visible over an area of about 16 acres.

After this success these bombs were used in many attacks and almost always there was photographic or other evidence of the devastation they had caused. In Berlin in September 1941 aerial reconnaissance disclosed a large area in the Lichtenburg district, probably demolished by one such bomb; the effect of its blast must certainly have damaged buildings far outside the area of total devastation, but such damage would scarcely have been visible from the air. In November 1941 there were reports of the terrible effects of these bombs in Berlin and of the fear which they inspired; when one dropped in the Nordhaven district twenty people were found dead in the street and many people actually sheltering in cellars were killed by the effect of blast alone. More than a hundred and fifty 4,000lb bombs were dropped on Düsseldorf on the night of 31 July/1 August 1942.

These new bombers and the new bombs were the main innovations of 1941. There seems to have been no correspondingly important innovation of policy in night bombing - though there were continual experiments in day bombing. It seemed to be only a question of increasing the force and regularly, throughout the spring, summer and autumn of 1941, the force increased; month by month there were larger numbers on the blackboards which hang on the walls of all the operation-rooms of Bomber Command. In spring and early summer the most obvious signs of the increase were that the major attacks, made when moonlight coincided with favourable weather got heavier and heavier. An attack on Berlin on 17/18 April was the heaviest the capital had yet had; on 8/9 May, in brilliant moonlight, a larger force was out over Germany than ever before, mainly to attack Hamburg and Bremen. But at the same time the average night's effort was increasing; the Rhineland industries and especially those of Mannheim were heavily attacked during May. The Ruhr was the most frequent target of June; preparations for the invasion of Russia had been noticed and Bomber Command did everything possible to hamper the movement of troops and supplies from the West to the East by attacking marshalling yards and railway towns; on 13 June, in particular, there was a most formidable attack on the communications of the Ruhr.

Endnotes Chapter 11

68   Squadron Leader John Henry Franks DFC was pilot of a Wellington on 57 Squadron when he was KIA on 29/30 June 1942.

Index